Diseases of Poultry
Their Etiology, Diagnosis, Treatment and Prevention

by Frank M. Surface

with an introduction by Jackson Chambers

This work contains material that was originally published in 1915.

This publication is within the Public Domain.

This edition is reprinted for educational purposes and in accordance with all applicable Federal Laws.

Introduction Copyright 2018 by Jackson Chambers

Self Reliance Books

Get more historic titles on animal and stock breeding, gardening and old fashioned skills by visiting us at:

http://selfreliancebooks.blogspot.com/

Introduction

I am pleased to present yet another title on Poultry.

The work is in the Public Domain and is re-printed here in accordance with Federal Laws.

As with all reprinted books of this age that are intended to perfectly reproduce the original edition, considerable pains and effort had to be undertaken to correct fading and sometimes outright damage to existing proofs of this title. At times, this task is quite monumental, requiring an almost total "rebuilding" of some pages from digital proofs of multiple copies. Despite this, imperfections still sometimes exist in the final proof and may detract from the visual appearance of the text.

I hope you enjoy reading this book as much as I enjoyed making it available to readers again.

Jackson Chambers

PREFACE

It is probably safe to say that considerably more than fifty per cent of the correspondence of those engaged in poultry work in the agricultural colleges and experiment stations in this country relates to poultry diseases. The poultryman or farmer sees that some, or perhaps all, of his birds are ill, and he straightway writes to the nearest college or station to know what the disease is, and what to do for it. The Maine Station has for many years been the recipient of a great number of such inquiries. It is an unfortunate, but in the nature of the case an unavoidable fact, that in many instances it is quite impossible to make any really satisfactory reply to these inquiries. In the vast majority of cases the person who writes the letter is quite untrained in pathology and either describes no symptoms at all or only those very general ones which are common to nearly all the ills of poultry. To one who has not handled this class of correspondence it would seem almost incredible that there should be so many letters of the following type: "I have lost about a third of my chickens in the last few days. They seem a little dumpish for a while and then die. What is the trouble and what shall I do for it?" To diagnose and prescribe on such a basis of information is impossible. Yet the hard fact remains that the correspondent's chickens *were* ill and *did* die, and he needs help to get out of the trouble.

Only recently has any attempt been made by our agricultural colleges to prepare its graduates in poultry husbandry and general agriculture to meet intelligently these

Preface

problems of poultry disease. Indeed even the veterinary schools have given but scant attention to avian pathology. Yet there exists, scattered in the literature, a large amount of definite information regarding poultry diseases. It was the purpose of the first edition of the circular on which the present book is based to give a clear and reasonably complete compilation and digest of the information then existing in the literature regarding the commoner diseases of poultry, their diagnosis, etiology, treatment, and prognosis.

The material was put together in the form which seemed most likely to meet the needs of the practical poultryman. It was soon found that in addition to this use, the book was coming to be rather widely employed as a text-book for courses in the diseases of poultry in agricultural colleges and high schools. In preparing the present edition the authors have kept more definitely in mind this second way in which the book may prove useful, and have added on that account some material not likely to be of immediate interest to the poultryman.

While none of the authors is engaged in researches in pathology, it is a fact that they have all had extensive experience in the appearance at autopsy of chickens dead from some one or more of a very wide variety of causes. There are but very few diseases mentioned in the book with which the authors have not had first-hand experience. It is hoped that the knowledge gained in this way will prove to have been of some critical value in the compilation of this book.

TABLE OF CONTENTS

	PAGE
PREFACE	v
LIST OF ILLUSTRATIONS	ix

CHAPTER

		PAGE
I.	GENERAL CONSIDERATIONS REGARDING THE TREATMENT OF POULTRY DISEASES	1
II.	POULTRY HYGIENE	8
III.	THE DIAGNOSIS OF THE DISEASES OF POULTRY	36
IV.	POULTRY MATERIA MEDICA	52
V.	DISEASES OF THE ALIMENTARY TRACT	57
VI.	POISONS	81
VII.	DISEASES OF THE LIVER	87
VIII.	FOWL CHOLERA, FOWL TYPHOID AND FOWL PLAGUE	102
IX.	TUBERCULOSIS	115
X.	INTERNAL PARASITES	133
XI.	DISEASES OF THE RESPIRATORY SYSTEM	147
XII.	DISEASES OF THE CIRCULATORY SYSTEM	182
XIII.	DISEASES OF THE NERVOUS SYSTEM	194
XIV.	DISEASES OF THE KIDNEYS, RHEUMATISM AND LIMBERNECK	199
XV.	EXTERNAL PARASITES	203
XVI.	DISEASES OF THE SKIN	233
XVII.	DISEASES OF THE REPRODUCTIVE ORGANS	245
XVIII.	WHITE DIARRHEA	283
XIX.	OTHER DISEASES OF CHICKENS	301
XX.	TUMORS	312
XXI.	POULTRY SURGERY	324
GLOSSARY OF TECHNICAL TERMS		329
INDEX		335

LIST OF ILLUSTRATIONS

FIGURE		PAGE
1.	Showing differences in constitutional vigor. (After H. R. Lewis)	5
2.	Silver-laced Wyandottes showing great constitutional vigor. (After D. S. Thompson)	7
3.	Curtain-front poultry house, exterior. (Original)	11
4.	Curtain-front poultry house, interior. (Original)	13
5.	Crematory. (Original)	33
6.	Skeleton of cock. (After Dürigen)	42
7.	Dissection of hen. (Original)	44
8.	Life history of a coccidium. (After Cole and Hadley)	73
9.	Showing condition of liver in blackhead. (After Moore)	95
10.	Heart blood of pigeon infected with fowl cholera. (After Kolle and Hetsch)	103
11.	Section of lung showing tubercle bacilli. (After Himmelberger)	117
12.	Breastbone of fowl showing emaciation in tuberculosis. (After Ward)	121
13.	Liver of fowl affected with tuberculosis. (After Ward)	122
14.	Spleen from tuberculous fowl. (After Koch and Rabinowitsch)	122
15.	Tuberculosis of intestines and mesenteries. (After Van Es and Schalk)	123
16.	Intestines and mesenteries of a fowl infected with tuberculosis. (After Ward)	124
17.	Syringe used in tuberculin test. (After Van Es and Schalk)	126
18.	Head of chicken showing positive tuberculin reaction. (After Van Es and Schalk)	127
19.	*Drepanidotænia infundibuliformis*, a tape worm of the fowl. (After Stiles)	135
20.	Intestine of fowl showing tape worms in nodular tæniasis. (After Pearson and Warren)	138
21.	Sketch showing method of introducing turpentine directly into crop. (After Gage and Opperman)	140
22.	Worms protruding from intestine of fowl. (After Bradshaw)	142
23.	*Heterakis perspicillum*. (After Salmon)	144

List of Illustrations

FIGURE		PAGE
24.	Trematode worm or fluke. (After Thompson)	145
25.	Lungs of a domestic fowl. (Original)	147
26.	Diagrammatic drawing of the left lung of a fowl. (Original)	148
27.	Ventral surface of right lung of a fowl. (From Gadow, after Stieda)	149
28.	Cross section of small air tubes of the lung of a goose. (From Oppel, after Schultz)	150
29.	Fowl's head showing infra-orbital tumor caused by roup. (After Roebuck)	157
30.	Showing appearance of a hen a day before death from roup. (After Harrison and Streit)	158
31.	Diphtheritic roup or canker. (After Roebuck)	166
32.	*Aspergillus fumigatus*. (After Mohler and Buckley)	176
33.	*Cytodites nudus*, the air-sac mite. (After Theobald)	181
34.	Bird suffering from *spirochætosis*. (After Kolle and Hetsch)	191
35.	Fowl *spirochætosis*. (From Kolle and Hetsch, after Barri)	192
36.	The common hen louse, *Menopon pallidum*. (From Banks)	206
37.	*Lipeurus variabilis*, a louse that infests poultry. (From Banks, after Denny)	207
38.	*Goniodes dissimilis*, a louse that infests poultry. (From Banks, after Denny)	207
39.	Feathers showing eggs or "nits" of the common hen louse. (Original)	208
40.	The common "red mite" of poultry, *Dermanyssus gallinæ*. (After Osborn)	214
41.	Normal leg and leg of hen affected with scaly leg. (After Mégnin)	217
42.	Leg of hen severely affected with scaly leg. (After Haiduk)	219
43.	Adult female of the mite *Knemidocoptes* (*Dermatoryctes*) *mutans*. (After Haiduk)	220
44.	Six-legged larva of *Knemidocoptes* (*Dermatoryctes*) *mutans*. (After Haiduk)	221
45.	Section of the skin of the leg of a fowl affected with scaly leg. (After Haiduk)	223
46.	Egg containing female of *Sarcoptes lævis* var. *gallinæ*. (After Theobald)	225
47.	*Symplectoptes cysticola*, connective tissue mite. (After Theobald)	227
48.	"Harvest bug," *Tetranychus* (*Leptus*) *autumnalis*. Larval form. (After Murray)	227

List of Illustrations

FIGURE		PAGE
49.	The poultry tick, *Argas persicus*, adult. (After Laurie)	228
50.	The poultry tick larva showing three pairs of legs. (After Laurie)	229
51.	The chicken flea, *Pulex gallinæ* or *avium*. (After Kaupp)	231
52.	Head and neck of fowl affected with generalized favus. (After Pearson)	233
53.	The fungus *Achorion schonleinii*, which causes favus in poultry	235
54.	Sore-head on comb, eyelids, and skin. (After Hadley and Beach)	239
55.	Reproductive organs of a hen. (After Duval)	247
56.	Showing a case of incomplete hermaphroditism. (Original)	252
57.	Oviduct removed from a laying hen and cut open along the point of attachment of the ventral ligament. (Original)	260
58.	Triple-yolked egg. (Original)	275
59.	Showing shapes of abnormal eggs sometimes found. (From von Durski, after Landois)	279
60.	The normal ovary of a laying hen. (After Rettger, Kirkpatrick, and Jones)	289
61.	Ovary from a hen infected with *B. pullorum*. (After Rettger, Kirkpatrick, and Jones)	290
62.	Showing how bacillary white diarrhea perpetuates itself in the breeding stock. (After Rettger and Stoneburn)	291
63.	Ten-day-old White Leghorn chicks showing symptoms of bacillary white diarrhea. (After Rettger and Stoneburn)	292
64.	Normal ten-day White Leghorn chicks. (After Rettger and Stoneburn)	293
65.	Trachea of a pheasant showing gape worms (*Syngamus trachealis*). (After Mégnin)	305
66.	A pair of *Syngamus trachealis* attached. (After Mégnin)	305
67.	A pair of *Syngamus trachealis*. (After Mégnin)	307
68.	Sarcoma chicken tumor No. 1. Second generation. (After Rous)	316
69.	Chicken tumor XVIII in the gizzard of the original fowl. (After Rous and Lange)	317
70.	Osteo-chrondrosarcoma produced by intramuscular injection of 4 cc. of the Berkfeld filtrate of an extract of chicken tumor No. VII. (After Rous and Murphy)	318
71.	The growth shown in figure 70 after it had been sawed open. (After Rous and Murphy)	319
72.	Cysto-adenoma on the serosa of the intestine. (After Pickens)	322

DISEASES OF POULTRY

CHAPTER I

General Considerations Regarding the Treatment of Poultry Diseases

There is general agreement on the part of authorities on poultry pathology and practical poultrymen of long experience that, in general, "doctoring" poultry is not advisable. The reasons for this attitude are primarily the following:

1. The unit of production with poultry (*i.e.*, the individual bird) is of relatively small value, and if a man's time is worth anything, it is too valuable to spend treating sick chickens individually unless they are show specimens of great individual value.

2. The "cured" chicken is a menace to the owner, because its identity is likely to be overlooked or forgotten, with the result that it goes into the breeding pen and perpetuates through its offspring the constitutional weakness which was one fundamental factor in bringing about the result that it, rather than some of its fellows, was ill.

This point of view has been well stated by Wright [1] in the following words:

"In a large proportion of cases of disease, the birds *ought* to die or be killed. Even where there is no constitutional taint, the fact that they have succumbed to circumstances which

---

[1] Wright, L., "The New Book of Poultry." London (Cassell & Company), 1905.

have not affected others, marks them out as the weakest, which unaided Nature would assuredly weed out, and which if we preserve and breed from, perpetuate some amount of that weakness in the progeny. Rheumatism, for instance, can be cured; of that there is no doubt. But the vast majority who have had such success, agree that the effects are either *never* recovered from, as regards strength and vigor, or else that the original weakness continues; and the same may be said of some severe contagious diseases, such as diphtheritic roup, which may affect the strongest. On the other hand, many diseases also apparently contagious, and so attacking healthy birds under certain predisposing conditions of exposure or other coincident strain upon the system, do not appear to leave serious results behind them, and are tolerably definite in symptoms and character. It is these which may be most successfully treated, and in which treatment is most worth while where fowls of value are concerned. But it is significant that nearly all breeders who rear really large numbers of poultry, gradually come to the conclusion that, except in special cases, with valuable birds, the most economical treatment of serious disease occurring in a yard is — execution. Concerning this matter each must judge for himself."

In the case of the utility poultryman, keeping poultry solely for the eggs and meat they produce, practically the only diseased conditions which it will pay him to treat at all are those in which the treatment can be applied to the flock as a whole, without the necessity of handling individual birds. Thus, for example, in cases where the flock "goes off its feed," or has simple indigestion or a mild cold, the birds can be treated successfully as a flock. On the other hand, in the case of the fancier, who has individual birds of considerable value, there will be a much wider range of diseases which he will feel that it is profitable for him to treat. There are, of course, certain diseased conditions which demand individual treatment, but in which the treatment is so simple and the outcome is almost certain to be so good, as to justify its employment even in the case of birds of ordinary value. An example of such a condition is found in a crop-bound bird.

Prevention Rather Than Cure the Ideal

The aim of every poultry keeper, whether his interest is in the fancy or the utility end of the business, should be to breed and manage his birds so as to prevent entirely, or reduce to a minimum, the occurrence of disease. In other words, the attitude should be that the end to be sought is to prevent the occurrence of disease, rather than to rely on a rather dubious ability to cure it after it is there. Such a standpoint is sound from every point of view; it is in line with the whole development of modern medicine. The poultry doctor should regard his function as the same as that of the Chinese physician, who is primarily employed to keep the patient from becoming ill, only secondarily to cure him.

Now there are fundamentally two factors involved in the continued maintenance of good health in poultry (or, for the matter of that, in any other animal). These are:

1. *A sound and vigorous constitution*, which if present, is something innate and "bred in the bone," and which, if *absent, must be bred into the stock.*

2. A system of *poultry management (including feeding, housing, etc.) which is thoroughly and absolutely hygienic.*

We shall consider each of these factors separately in some detail.

Breeding for Health, Vigor and Sound Constitution

To have a high degree of constitutional vigor in the foundation stock is one of the most certain assurances that the poultryman will not be troubled with disease. This is of primary importance. In order to breed constitutional vigor into the flock, the poultryman must train himself to recognize at a glance the condition of his birds. Are they in good condition or not? Regarding the aspect of fowls in health and disease Salmon [1] has the following to say:

"We say that a bird is in good health when it appears lively, has a clear eye, a bright red comb, is quick and active in its movements, has a good appetite and when the various organs perform their functions in the manner in which they are observed to act in all birds that are vigorous and thriving. On the other hand, we say a bird is diseased when some function or functions of its body are not performed as they are in the great majority of individuals, or when some organ presents an unusual form or appearance. Disease has, therefore, been defined as a life the manifestations of which deviate more or less from the normal. Practically, we say a bird is diseased when we observe that one or more of its functions are not carried on in a normal manner, or when we find unusual growths, injuries, or parasites affecting any of its organs."

Having acquired the ability to see the individual birds as individuals, the next step is to learn to distinguish a good bird from a poor one. Here it is ever to be kept in mind that the primary and most essential characteristic of a good bird must always be a sound constitution and plenty of vitality and vigor. Without these qualities it is impossible to have first class stock. Constitutional vigor and vitality may be put as a fundamental requisite in the successful practical breeding of poultry. In all kinds of breeding operations whether for utility purposes, or for the fancier's show pen, or for the purpose of experimentation in the field

[1] Salmon, D. E., "The Diseases of Poultry." Washington (no date).

FIG. 1. — Chicks eight weeks old, hatched and reared at the New Jersey Agricultural Experiment Station, showing marked differences in constitutional vigor and vitality at that age. The bird in the center is a strong, vigorous, well-developed animal. The one at the right is markedly lacking in vitality, while the one at the left is a worthless weakling, which will never amount to anything and might as well be killed at once. (After H. R. Lewis.)

of heredity, the first selection of birds for the breeding pen should be made on the basis of their general constitutional vigor. No bird which shows signs of weakness in this fundamental regard should ever be used as a breeder under any circumstances. If such a bird is used the breeder will eventually have to pay the penalty.

The external, visible evidences of a sound constitution and a possession of abundant vitality and vigor are numerous. In the first place the bird of sound constitution will be in perfect health. Perhaps its most striking characteristic will be an independence of disposition and demeanor. By this is not necessarily meant aggressiveness. The bird, whether male or female, which is forever picking quarrels with its fellows is by no means always the bird of greatest vigor. Strange as it may seem a bird may indeed be very far from a mollycoddle and yet have a peaceable disposition. It may be taken as an unfailing characteristic of birds of high constitutional vigor, however, that they are able to take care of themselves and may not be imposed upon, or bullied by their fellows, with impunity. While they may not pick a quarrel, they are abundantly able to make a forceful presentation of the merits of their end of any debate which another bird may choose to enter upon with them. In other words they have, as has been said, an independence of disposition; an ability, reaching to the limits of gallinaceous capacity, to meet all situations which may arise in the day's work of a fowl, whether food getting, fighting, rearing young, or what not.

The bird of high constitutional vigor will have a thrifty appearance, with a bright eye, and clean, well-kept plumage. The head will be broad and relatively short, giving in its appearance plain indication of strength. It will show nothing of the long-drawn-out, sickly, crow-like appearance of the head which is all too common amongst the inhabitants of the average poultry yard. The beak will be relatively short and strong, thus correlating with the general conformation of the head. Comb and wattles will be bright in color and present a full-blooded, healthy, vigorous appearance.

The body of the bird of high constitutional vigor will be broad and deep and well meated, with a frame well knit together, strong in the bone but not coarse. In fowls of strong

Fig. 2. — Six Silver-laced Wyandottes, which have great constitutional vigor and vitality. These birds averaged to lay 204 eggs each in a year. Only birds in perfect health and high constitutional vigor can make such records. (After D. S. Thompson.)

constitution and great vigor all the secondary sexual differences will usually be well marked. In other words the males will be masculine to a degree in appearance and behavior, and the females correspondingly feminine. It must be noted, however, that this last is a general rule to which there are occasional exceptions.

CHAPTER II

POULTRY HYGIENE

SECOND in importance only to high constitutional vigor and health is attention to the basic rules of hygiene and sanitation in the management of poultry. In view of the prevalent misunderstanding or lack of understanding of these principles it seems wise to devote one chapter to an outline of the more important points which need to be looked after in hygienic poultry keeping. Attention to the rules and principles here set forth will go a great way towards preventing the occurrence of disease. This does not mean that if these rules are *not* followed disease and destruction will forthwith result. Every one knows of plenty of instances of more or less successful poultry keeping under the most insanitary and unhygienic of conditions. So, similarly, human beings are able, when forced to do so, to live under unhygienic conditions. But every civilized country in the world believes that the most economical insurance against the steady loss of national wealth which the prevalence of disease involves is the enforcement of sanitary regulations throughout its domain. Again, many men who do not carry fire insurance on their buildings go through life without having any of them burn down. But this is no argument against the fact that it is a sound economic policy to carry fire insurance. In poultry keeping many may be successful for a time in managing their birds in defiance of the laws of sanitation and hygiene; *a very few* may be successful in this practice for a long time, but in the long run the vast majority will find that thorough, careful and intelligent attention to these laws will be one of the best guaranties of *permanent* success that they can find.

Poultry hygiene and sanitation will be considered here under seven main heads, as follows: I. Housing. II. Feeding. III. The Land. IV. Exercise. V. External Parasites. VI. Disposal of the Dead. VII. Isolation of Sickness. What is said under all of these heads is intended to apply (unless a specific statement to the contrary is made) both to adult birds and to chicks. No discussion of the hygiene of incubation, or of the relative merits of artificially and naturally hatched chickens will be undertaken here, because these are special subjects falling outside the field of general poultry hygiene.

I. Poultry House Hygiene and Sanitation

A. General Principles of Poultry Housing. — In the management of adult fowls there are in the main two things to be considered, housing and feeding. A vast multitude of methods of doing these two things to poultry have been tried during the history of the industry.

There have been published plans for poultry houses of all conceivable shapes and sizes. Long houses, short houses, tall houses, low houses; square, hexagonal, octagonal and round houses; heated houses and cold houses; all these and many more have had their advocates, and detailed plans for their construction can be found. It would appear that there must be realized here the primary condition of the experimental method, namely the "trying of all things." It only remains to discover that which is "good" in order that we may "hold fast" to it.

This discovery had indeed been made in regard to a few

of the basic things in the housing of poultry. It would be strange if something had not come out of all the indignities to which innocent and inoffensive generations of fowls have been submitted in the way of dwelling accommodations. It is now clearly recognized, and generally admitted by all competent poultrymen, that certain things are absolutely essential in any poultry house which is to give good results. These are (1) fresh air, (2) freedom from dampness, (3) freedom from draughts, (4) sunlight and (5) cleanliness.

If these five things are realized in a poultry house the birds will thrive and be productive in it, provided they are well and regularly fed and watered. It makes no difference particularly to the well-being of the birds how these necessary specifications of their dwelling are attained. To the poultryman, however, it is important that they be attained at the smallest expense, having regard to (a) initial cost, (b) repairs and up-keep and (c) labor necessary to operate the house to get the specified results. The housing problem is to the poultryman, then, both a biological and an economic one. The biological solution is definite. The requisites named above must be met, and there is one additional factor to be taken into account; namely, size of house. Experiments made at various times and places indicate clearly that in northern climates, where birds must be shut up in the house during a part of the year in order to give best results, there should be allowed in the house at least three square feet of floor space per bird, and preferably a little more. Four square feet floor space per bird is a liberal allowance.

A factor which it was formerly thought necessary to control in the housing of poultry was the temperature. It was long held that if fowls were to lay well in the winter it was necessary that they should be in a heated house. Later experience has shown conclusively that this was an utterly

FIG. 3. — Curtain-front poultry house No. 3. at the Maine Agricultural Experiment Station. This type of house has proven very satisfactory in cold northern climates.

fallacious idea. As a matter of fact, even in the coldest climates, fowls will lay better during the winter months in a properly constructed house wide open to the outside air in the day time, so that they are living practically out of doors, than in any heated house which has yet been devised. If a laying house is *dry* the temperature factor may be neglected. If a house has a tendency to dampness, it will give poor results regardless of temperature.

From the economic standpoint there are two systems of housing poultry to be considered. One of these is the system of long, continuous houses for the laying birds. The other is the so-called colony house system, in which the birds are housed in small separate houses which may either be set a considerable distance apart over a relatively wide area, or may be placed relatively near one another. Each system has its strenuous advocates. Experience covering a fairly long period of years now has demonstrated that both systems have good points. As to which shall be adopted in a particular instance depends upon a variety of considerations, each in some degree peculiar to the particular case in hand.

In the extreme northern part of the country where the climate is very cold in the winter and there is an abundance of snow there can be no question that the long house is much to be preferred to a colony system. There are two reasons for this. In the first place experience indicates that the birds are somewhat more productive and keep in better condition in a properly constructed and managed long house than in colony houses. Furthermore the labor expense involved in caring for a given number of fowls is much less, under such climatic conditions, than with the colony house system, where the birds are scattered over a wider area and more paths must be broken out in the snow.

The great advantage of the colony house system is its

Fig. 4. — Interior of curtain-front poultry house No. 3. This shows the trapnests at the end, feeding trough in front, water bucket in center and roosting boards in rear.

flexibility. Furthermore it gets around the troubles involved in the contamination of the ground by the long-continued keeping of poultry on the same small area. In general, local conditions and circumstances must decide in each individual case which system of housing shall be adopted.

B. Cleanliness. — The thing of paramount importance in the hygienic housing of poultry is *cleanliness*. By this is meant not merely plain, ordinary cleaning up, in the housewife sense, but also bacteriological cleaning up; that is, *disinfection*. All buildings or structures of whatever kind in which poultry are housed during any part of their lives should be subjected to a most thorough and searching cleaning and disinfection *at least* once every year. This cleaning up should naturally come for each different structure (*i.e.*, laying, colony or brooder house, individual brooder, incubator, etc.) at a time which just precedes the putting of new stock into this structure.

How to clean a poultry house. — Not every poultryman of experience even, knows how *really* to clean a poultry house. The first thing to do is to remove all the litter and loose dirt which can be shoveled out. Then give the house — floor, walls and ceiling — a thorough sweeping and shovel out the accumulated débris. Then play a garden hose, with the maximum water pressure which can be obtained, upon floor, roosting boards, walls and ceiling, until all the dirt which washes down easily is disposed of. Then take a heavy hoe or roost board scraper and proceed to scrape the floor and roosting boards, *clean* of the trampled and caked manure and dirt. Then shovel out what has been accumulated and get the hose into action once more and wash the whole place down again thoroughly and follow this with another scraping. With a stiff bristled broom thoroughly scrub walls, floors, nest boxes, roost boards, etc. Then after another rinsing down and cleaning out of accumulated dirt, let the house dry out for a day or two. Then make a searching inspection to see if any dirt can be discovered. If so, apply the appropriate treatment as outlined above. If, however, everything *appears* to be clean, the time has come to make it *really* clean by *disinfection*. To do this it is necessary to spray or thoroughly wash with a scrub brush wet in the solution used all parts of the house with a good disinfectant *at least twice*, allowing time between for it to dry. For this purpose 3 per cent cresol solution is recommended. The chief thing is to use an effective disinfectant and plenty of it, and apply it at least twice. A discussion of disinfectants immediately follows this section. To complete the cleaning of the house, after the second spraying of disinfectant is dry apply a liquid lice killer (made by putting 1 part crude carbolic acid or cresol with 3 parts kerosene) liberally to nests and roosts and nearby walls. After all this is done the house will be *clean*. In houses cleaned annually in this way the first step is taken towards hygienic poultry keeping.

The same principles which have been here brought out should be applied in cleaning brooders, brooder houses, and other things on the plant with which the birds come in contact.

What has been said has reference primarily to the annual or semiannual cleaning. It should not be understood by this that no cleaning is to be done at any other time. On the contrary the rule should be to keep the poultry house *clean* at all times, never allowing filth of any kind to accumulate and using plenty of disinfectant.

Disinfection. — In the matter of disinfection there are several options open to the poultryman. He may make his own disinfectant, or he may purchase proprietary compounds like Zenoleum, Carbolineum or a host of other "eums" which confront him at every turn in his reading of poultry periodicals, or he may buy a plain disinfectant like formaldehyde, or carbolic acid.

There is no more effective general disinfectant than formaldehyde, and it also has the advantage of being cheap. We have used it regularly for some years past with excellent results. A 5 per cent solution of commercial formalin in water is applied to walls, floors and roost boards by means of a pressure spray pump. Various hand pumps of this type are on the market. It will pay the poultryman to get one of the well made higher priced sorts. If a spray pump is not available the formaldehyde solution may be put on with a brush. In any case a liberal amount should be used. When applying it all doors and windows should be open to diminish as much as possible the irritating effect of the vapor on the worker. His hands should also be protected by the use of well oiled leather gloves.

Some writers have advocated the formaldehyde gas method for disinfecting poultry houses, using the permanganate method of generating. This, however, is indicated only for rooms which can easily be closed up air tight. It costs too much in time and trouble to make any form of "fresh air" poultry house even moderately air tight. The formaldehyde gas method is well adapted to disinfecting and fumigating feed rooms, incubator cellars, brooder houses and all houses which can be readily made air tight. For the benefit of those who wish to use the method for such purposes the following directions are given. This will give a very strong fumigation and disinfection but such is indicated about poultry establishments.

Formaldehyde gas disinfection. — First make the room as tight as possible by stopping cracks, key-holes, etc., with pieces of cloth or similar substance. Open bins and doors of closets, etc., to allow free access of the gas. Use a metal or earthen dish for a generator, of sufficient size so that the liquid will not spatter or boil over on the floor, since the permanganate will stain. The temperature of the room should not be below 50° F. and more effective disinfection will be obtained if the temperature is 80° F. or above at the beginning. Sprinkle boiling water on the floor or place a kettle of boiling water in the room to create a moist atmosphere. Spread the permanganate evenly over the bottom of the dish and quickly pour in the formaldehyde (40 per cent strength as purchased). Leave and tightly close the room at once and allow to remain closed for 4–6 hours or longer, then air thoroughly. *Use 23 ounces of permanganate and 3 pints of formaldehyde to each 1000 cubic feet of space.*

Cresol disinfectant.— For a disinfectant of the coal-tar or carbolic acid type, we have found a cheap and satisfactory sort to be compound cresol solution. This may be used alone or as a second spray following formalin for spraying and disinfecting the houses after they are cleaned, disinfecting brooders, brooder houses, incubators, nests and everything else about the plant which can be disinfected with a liquid substance. It is particularly effective against mites and other insect pests. It has been very satisfactory in disinfecting incubators between hatches. Any person can easily make this disinfectant. The following directions for its manufacture are quoted from Bulletin 179 of the Maine Agricultural Experiment Station.

The active base of cresol soap disinfecting solution is commercial cresol. This is a thick, sirupy fluid, varying in color in different lots from a nearly colorless fluid to a dark brown. It does not mix readily with water, and, therefore, in order to make satisfactory a dilute solution, it is necessary first to incorporate the cresol with some substance like soap which will mix with water and will carry the cresol over into the mixture. The commercial cresol, as it is obtained, is a corrosive substance, being in this respect not unlike carbolic acid. It should, of course, be handled with great care and the pure cresol should not be allowed to come in contact with the skin. If it does so accidentally the spot should be immediately washed off with plenty of clean water. The price of commercial cresol varies with the drug market. It can be obtained through any druggist. In purchasing this article one should order simply "commercial cresol."

Measure out $3\frac{1}{5}$ quarts of raw linseed oil in a 4 or 5 gallon stone crock; then weigh out in a dish 1 lb. 6 oz. of commercial lye or "Babbit's potash." Dissolve this lye in as little water as will completely dissolve it. Start with $\frac{1}{2}$ pint of water, and if this will not dissolve all the lye, add more water slowly. Let this stand for at least 3 hours until the lye is completely dissolved and the solution is cold; then add the *cold* lye solution very slowly to the linseed oil, stirring constantly. Not less than 5 minutes should be taken for the adding of this solution of lye to the oil. After the lye is added continue the stirring until the mixture is in the condition and has the texture of a smooth homogeneous liquid soap. This ought not to take more than a half hour. Then while the soap is in this liquid state, and before it has a chance to harden add, with constant stirring, $8\frac{1}{2}$ quarts of commercial cresol. The cresol will blend perfectly with the soap solution and made a clear, dark brown fluid. The resulting solution will mix in any proportion with water and yield a clear solution.

Cresol soap is an extremely powerful disinfectant. In the Station poultry plant for general purposes of disinfecting the houses, brooder houses, incubators, nests and other wood work, it should be used in a 3 per cent solution with water. Two or three tablespoons of the cresol soap to each gallon of water will make a satisfactory solution. This solution may be applied through any kind of spray pump or with a brush. Being a clear, watery fluid it can be used in any spray pump without difficulty. For disinfecting brooders or incubators which there is reason to believe have been particularly liable to infection with the germs of white diarrhea or other diseases the cresol may be used in double the strength given above and applied with a scrub brush in addition to the spray.

C. Fresh Air and Light. — Too great stress cannot be laid on the importance of plenty of fresh air in the poultry house if the birds are to keep in good condition. And it must be remembered in this connection that "fresh" air, and cold stagnant air are two very different things. Too many of the types of curtain front and so-called "fresh" air houses now in use are without any provision other than an obliging southerly wind, to insure the circulation or changing of air within the house. Even with an open front house it is wise to provide for a *circulation* of air in such way that direct drafts cannot strike the birds. This applies not only to the housing of adult birds in laying houses, but also to the case of young stock in colony houses on the range. Further a circulation of fresh air under the hover in artificial rearing is greatly to be desired and will have a marked effect on the health and vigor of the chicks.

Not only should the poultry house be such as to furnish plenty of fresh air, but it should also be *light*. The prime importance of sunlight in sanitation is universally recognized by medical authorities. Disease germs cannot stand prolonged exposure to the direct rays of the sun. Sunlight is Nature's great disinfectant. Its importance is no less in poultry than in human sanitation. The following statement made some years ago (1904) by a writer signing himself "M" in *Farm Poultry* (Vol. 15) brings home in a few words the importance of having plenty of light in the poultry house.

"Light in the poultry house has been found by a writer a *great help in keeping the house clean and keeping the fowls healthy.* Probably there is no greater assistance to the diseases of poultry than dark and damp houses, and dark houses are frequently damp. In recent years I have had both kinds of experience, those with the hens confined in a large, dry and light house, and with hens confined in a dark house in which a single window looking towards the setting sun furnished the only light. Being forced to use the latter building for an entire winter I found it impossible to get it thoroughly dried out after a rain had rendered the walls damp. By spring some of the fowls that had been confined there began to die of a mysterious disease and a post-mortem examination showed it to be liver disease. Later the roup broke out in the same house and this dread disease continued with the flock for months exacting a heavy toll in laying hens."

D. Avoid Dampness.—Of all unfavorable environmental conditions into which poultry may, by bad management, be brought, a damp house is probably the worst. Nothing will diminish the productivity of a flock so quickly and surely as will dampness in the house, and nothing is so certain and speedy an excitant to roup and kindred ills. *The place where poultry are housed must be kept dry if the flock is to be productive and free from disease.*

E. Provide Clean and Dry Litter.—Experience has demonstrated that the best way in which to give fowls exercise during the winter months in which, in northern climates at least, they must be housed the greater part if not all of the time, is by providing a deep litter in which the birds scratch for their dry grain ration. For this litter the Maine Agricultural Experiment Station uses pine planer shavings, with a layer of straw on top. Whatever the litter it should be changed as often as it gets damp or dirty.

II. Hygienic Feeding

Having housed our fowls they must be fed. Here the same sort of history is to be found as in the case of housing. Substantially all known edible substances must, at some time or other, have been suggested or tried as component parts of the rations of fowls. Not only have many and curious substances been suggested as poultry food, but they have been combined in formulæ as weird as a medieval apothecary's prescription. Actually practical poultry feeding is much more of an art than a science, in the present state of knowledge. While for pedagogical reasons it seems wise in the teaching of poultry husbandry to spend a considerable amount of time in calculating balanced rations and nutritive ratios, it is very doubtful if all such activity has any real or tangible relation to practical poultry feeding.

Such attempts at a science of poultry feeding would appear to suffer from a serious defect. The assumption is made in calculating a nicely balanced ration that all hens are going to partake of this ration in the same way. But this is very far from the biological actuality. Some individual hens like no grain except corn, and if fed a mixture will eat only corn. Others are very partial to beef scrap, and so on. To any one who studies the behavior of fowls it is clear that the ration on paper and the ration in the crop are two very different things.

The successful feeding of poultry depends upon experience and acquaintance with fowls. The basic biological factor is, once more, individuality. Each individual hen is an independent living thing, possessing well marked likes and dislikes of her own with respect to food. There can be no question that the best results in the way of egg production and meat production would be obtained if a skillful feeder could feed each individual fowl by and for itself.

Evidence that this is the case is found in the fact, which is universal wherever poultry is kept, that on the average fowls kept in small flocks, of, say, under 25 birds each, do relatively much better than larger flocks. The production and money returns per bird are greater. The fundamental reason for this is that the birds in small flocks get better care as individuals. When a man has only such a small number to take care of he can recognize their individual peculiarities more easily. Furthermore an individual bird stands a better chance of having its peculiar taste gratified in a small than in a large flock.

So while the biological ideal would be to feed each bird individually, this is obviously impossible in practice. With poultry the individual unit of production (the hen) is so small that it must be handled in flocks. The correct principle of management is to feed and handle a flock in such a way as to afford the maximum opportunity for the expression and gratification of the individual preferences of the component units, with a minimum labor cost. The larger the flock and the plant as a whole, the more machine-like the methods of feeding and handling must be. They must of necessity be calculated to suit that mythical creature, the average hen. Coincidently the total production or profit per bird will diminish. Presently a point is reached in size of plant where the outgo exceeds the income over a period of years. Such a plant if it has a hustling business man at the head takes a fancy name to itself, advertises a great deal, invents a "system," writes and sells a book about it, manufactures incubators and supplies, in general endeavors to make a loud noise about what a profitable thing the poultry business is, and finally goes dismally, completely and permanently "broke."

In the practical feeding of flocks of poultry large enough to be a commercial proposition, the methods which have been worked out empirically by the successful poultryman are essentially attempts to satisfy the individual tastes of the birds to as great a degree as possible, at a minimum labor cost. This result is obtained in practice by offering to the flock a variety of food materials so that they may have some opportunity of choice as to what they shall eat. If we feed corn, wheat, and oats the fowl which likes corn has the opportunity to live on corn, whereas the fowl which likes about three parts wheat and one part oats is able to satisfy her taste in this regard.

As a result of this manifest need for a variety of food it has come about that the practice now generally accepted as best is to put regularly before fowls food substances belonging to four different categories. These categories are:

1. Dry whole (or coarsely broken) grains (*e.g.*, corn, wheat, oats, barley, etc.).

2. Ground grains (*e.g.*, bran, middlings, corn meal, linseed meal and other finely ground grains).

3. Animal products (*e.g.*, beef scrap, blood meal, fish scrap, green cut bone, etc.).

4. Succulent or green foods (*e.g.*, mangolds, cabbages, beets, sprouted oats, green corn fodder, etc.).

The proportions in which these different kinds of food material are fed differ to a considerable extent among different poultrymen. The *exact* proportions in which they are given really matter very little, owing to the fact, already brought out, that the hen compounds her own ration to her own taste if given the material. Furthermore it makes little difference whether the ground grains are fed dry or wet. It is cheaper to feed them dry (because of labor saved), and therefore the "dry-mash system" of feeding has become popular.

There are certain basic principles of hygienic feeding which must always be looked after if one is to avoid diseases. There are:

A. Purity. — It should be a rule of every poultryman never to feed any material which is not clean and wholesome. Musty and moldy grain, tainted meat scraps or cut bone, table scraps which have spoiled, and decayed fruits or vegetables should never be fed. If this consideration were always kept in mind many cases of undiagnosed sickness and deaths, and low condition in the stock would be avoided. Keep all utensils in which food is placed *clean.*

B. Avoid Overfeeding. — Intensive poultry keeping involves of necessity heavy feeding, but one should constantly be on the lookout to guard against overfeeding, which puts the bird into a state of lowered vitality in which its natural powers of resistance to all forms of infectious and other diseases are reduced. The feeding of high protein concentrates like linseed or cotton seed meal needs to be particularly carefully watched in this respect.

C. Provide Plenty of Green Food. Under natural conditions poultry are free eaters of green grass and other plants. Such green food supplies a definite need in metabolism, the place of which can be taken by no other sort of food material. It is not enough merely to supply *succulence* in the ration. Fowls need a certain amount of succulent food, but they also need *fresh green food.* Green sprouted oats, when properly prepared, are an excellent source of winter green food. Full directions for sprouting oats are given in Bulletin 179 of the Maine Station.

D. Provide Fresh and Clean Drinking Water. — The most sure and rapid method by which infectious diseases of all kinds are transmitted through a flock of birds is by means of the water pail from which they all drink in common. Furthermore the water itself may come from a contaminated source and be the origin of infection to the flock. Finally it is difficult to devise any satisfactory drinking fountain in which the water is not liable to contamination from litter, manure, etc. All these considerations indicate the advisability of adding to all drinking water which is given to poultry some substance which shall act as a harmless antiseptic. The best of all such substances yet discovered for use with poultry is potassium permanganate. This is a dark, reddish-purple crystalline substance which can be bought of any druggist. A pound will last for a long time. It should be used in the following way: In the bottom of a large mouthed jar, bottle or can, put a layer of potassium permanganate crystals an inch thick. Fill up the receptacle with water. This water will dissolve all of the crystals that it is able to. This will make a stock saturated solution. As this solution is used add more water and more crystals as needed, always aiming to keep a layer of undissolved crystals at the bottom. Keep a dish of stock solution like this alongside the faucet or pump where the water is drawn for the poultry. *Whenever any water is drawn for either chicks or adult fowls add enough of the stock solution to give the water a rather deep wine color.* This means 1 to 2 teaspoons of the stock solution to 10 quarts of water. At the same time one should clean and disinfect the drinking pails and fountains regularly, just as he would if he were not using potassium permanganate. At the Maine Station plant for some years past no bird has ever had a drink of water from the time it was hatched which did not contain potassium permanganate, except such water as it got from mud puddles and the like.

Dr. G. B. Morse,[1] a well known authority on poultry diseases, had the following to say regarding this point in a recent address. After describing the potassium permanganate method, as well as two others, directed to the same end,

[1] Morse, G. B., "The Gospel of Cleanliness of Poultrymen." *Reliable Poult. Jour.*, Vol. 17, No. 8, pp. 756, 757, 775–777, 1910.

but in the opinion of the present writers not so desirable as this, he went on to say: "Water-borne diseases are frequent in the poultry yard. Clean and disinfect your drinking-fountains (and you must) ever so well, if you are permitting, consciously or unwittingly, to run at large one bird sick with any of the contagious diseases of the head parts or with bowel diseases, you may count on that water supply being contaminated in less than one hour's time. In the case of a large flock affected with flagellate diarrhea I have myself found the flagellates in less than one hour's time in the drinking water which had been sterilized and placed in thoroughly disinfected fountains. Do you not see where such a condition as this forces you? Right up against the principle of the individual drinking cup. Ridiculous, do you say? Not a bit. I did not say 'the individual drinking cup,' but the 'principle of the individual drinking cup.' Boards of health are recognizing that by means of the common, public drinking cup foul and terrible diseases are being spread among people. It is just so with your poultry, and while you cannot adopt the individual cup you can incorporate the principle of it in your hygienic methods by adding . . . one of the antiseptics named. It is true, in the proportions named, these remedies do not disinfect the water, only act as antiseptics, that is, act to hinder the developement of bacteria and other microbes. The water itself should be changed frequently. This hindering of microbian growth occurs not only in the fountain but is kept up in the intestinal tract."

III. The Land

One of the most important considerations in poultry sanitation is to keep the ground on which the birds are to live, both as chicks and as adults, from becoming foul and contaminated. This is not a very difficult thing to do if one has enough land and practices a definite and systematic crop rotation in which poultry form one element. On the open range where chicks are raised a four year rotation is operated at the Maine Agricultural Experiment Station and serves its purpose well. This system of cropping for the shorter period is as follows: First year, chickens; second year, a hoed crop, such as beets, cabbage, mangolds or corn; third year, seed down to timothy and clover, using oats or barley as a nurse crop; fourth year, chickens again. When the land can be spared it is left in grass the fourth year, and the chickens are not put on it until the fifth year. The reason for the particular crops mentioned above being used is that they are all things which can be very advantageously used in furnishing green food for the poultry at different seasons of the year.

To maintain the runs connected with a permanent poultry house, where adult birds are kept, in a sweet and clean condition is a more difficult problem. About the best that one can do here is to arrange alternate sets of runs so that one set may be used one year and the other set the next, purifying the soil so far as may be by plowing and harrowing thoroughly annually, and planting exhaustive crops. Failing the possibility of alternating in this way, disinfection and frequent plowing are the only resources left.

The following excellent advice on this subject is given by the English poultry expert Mr. E. T. Brown[1]: "Tainted ground is responsible for many of the diseases from which fowls suffer, and yet it is a question that rarely receives the attention it deserves. The chief danger of tainted soil arises when fowls are kept in confinement, but still we often find that even with those at liberty the land over which they are running is far from pure. So long as the grass can be kept growing strongly and vigorously there is small fear of

[1] *Farm Poultry*, Vol. 18.

foul ground, as the growth absorbs the manure; it is when the grass becomes worn away that the chief danger arises. The manure constantly falling upon the same small area, and there being nothing to use it up, the land is bound in a short space of time to become so permeated as to be thoroughly unfit for fowls. The question is very often asked in connection with this subject as to how many fowls a certain sized piece of land will accommodate the whole year through. Occasionally one may see in some of the agricultural or poultry journals this question answered, but as a matter of fact to give any stated number is most misleading. It depends very largely upon the class of soil, as some can carry twice as many birds as others; it depends upon the breed of poultry, some being much more active than others, and thus requiring more space; it depends, too, upon the time of year, because during the spring and summer, when there is an abundance of vegetable growth in the soil, a considerably larger number of birds can be maintained than during the autumn or winter. The number must be varied according to these circumstances, and no hard and fast rule is applicable."

"The results of tainted ground are generally quickly noticeable, as the fowls have a sickly appearance, the feathers lose their brilliant luster, and the wings begin to droop. Roup, gapes, and other ailments speedily show themselves, causing, if not death itself, considerable loss and unpleasantness. One of the greatest advantages to be derived from portable houses is that they so greatly reduce the risk of tainted ground, as they are being constantly moved from one place to another, thus evenly distributing the manure. When it is remembered that each adult fowl drops nearly a hundredweight of manure in the course of a year, the importance of this question will be immediately realized. It is quite possible, however, provided that suitable precautions are taken, to keep a comparatively small run pure for a long time. If the grass is short it should be occasionally swept, in this manner removing a good deal of the manure. Another important point is to always have around the house a space of gravel, upon which the birds should be fed, and if swept once or twice a week this will have a wonderful effect in preserving the purity of the grass portion. Any one who has observed poultry will know how fond they are of constantly being near the house, and thus the greater portion of their droppings falls within its immediate vicinity. The shape of the run also has a great bearing upon the length of time it will remain untainted, a long narrow run being much superior to a square one. I have proved by my own experience how true this is, and probably a long and narrow run, containing the same amount of space will remain pure twice as long. It is unnecessary here to go into a full explanation of why this is so, but I may state the fact, which I am confident is quite correct. If the space at one's disposal is very limited it is a good plan to divide it into two equal parts, placing the house in the middle. During one year one-half would be available for the fowls, the other being planted with some quickly growing vegetables, the order being reversed the year following. The vegetable growth has the effect of quickly using up the manure, and in this manner quite a small plot of land can be heavily stocked with poultry for an unlimited number of years. If the soil becomes at all foul it is a good plan to water it with a 1 per cent solution of sulphuric acid, or to apply a light dressing of gas lime."

IV. Exercise

If poultry are to be in good condition, and maintain their normal resistance to disease *they must exercise*. As chicks they will do this on the range. In the case of adults (in cold climates) the most feasible way to bring this about is to provide litter and make the birds scratch for their food.

V. External Parasites

In hygienic poultry keeping the birds must be kept reasonably free at all times of lice, mites, and all other forms of external parasites. Directions for dealing with this matter are given in detail farther on in this book in the chapter on External Parasites. It is desired here merely to call attention to the matter as one of general principles of hygienic poultry management.

VI. Disposal of Dead Birds

On every poultry plant and around every farm there are bound to occur from time to time a greater or less number of deaths of chickens and adult fowls from disease or other natural causes. The disposal of these dead bodies offers a problem to the poultryman, the correct solution of which may in many cases become a very important matter. This is especially true in the cases of death from contagious diseases, which include a considerable proportion of the deaths of poultry generally. The method usually practiced by the farmer and poultryman for the disposition of dead carcasses is unsanitary in the extreme. To throw the dead bodies on the manure pile is to invite the spread of disease on the plant. Burying is far from being a satisfactory way of dealing with the matter for two reasons. Unless the grave is dug deep, which costs a good deal of time and labor, there is considerable likelihood that dogs or other marauding animals will dig out the carcasses, and, after feeding on them, scatter the remains around on the top of the ground.

Furthermore, burying cannot be resorted to at all during the winter months when the ground is frozen.

The only really sanitary method of dealing with dead bodies is to incinerate them. The difficulty of following this plan in practice is that the farmer or poultryman usually does not have any suitable source of heat ready at hand at all times. To be sure, during certain seasons of the year, those poultrymen who employ large brooder houses with a hot water heating system have a furnace in operation, and the dead chicks can be burned up in the furnace. This, however, covers only a part of the year. At other times resort must be had to burying or some other means of disposal, as the poultryman is not likely to fire up a large furnace for the sake of burning a few dead birds.

To meet this requirement there has recently been devised at the Maine Agricultural Experiment Station the small crematory here described. The construction was carried out with the idea of keeping the first cost as low as possible, in order that there should be nothing about it which any poultryman or farmer could not easily afford to duplicate. As a matter of fact, the cost of materials for the crematory was less than ten dollars. The labor was done by the poultryman and his assistant at odd times, when an hour or two could be spared for this work. The result is, therefore, not beyond the reach of any poultryman or farmer. At the same time the crematory is so satisfactory in operation that any one who builds one will wonder, after he has completed and used it for a time, why he did not long before have so simple and sanitary an adjunct to his plant.

The crematory shown in Fig. 5 is very simple in construction. It consists essentially of a cement base or fire box, bearing on its top a series of grate bars which are in turn covered by a cremating box or oven in which the material to be incinerated is placed.

The crematory is sufficiently large to take care of all the needs of a plant carrying 1000 head of adult stock, raising 3000 to 4000 chickens annually, and in which a good deal of anatomical and physiological research is going on, necessitating a much larger amount of waste animal material than the ordinary commercial poultryman would have. Therefore, it is doubtful if it would be necessary in any but the very largest commercial plants to build a larger crematory than the one here described.

In building this an excavation was first made for the base, in which a lot of loose stones and gravel were placed, in order to secure adequate drainage below the cement. On top of this the cement base and fire box were made.

This base consists essentially of a rectangular box made of cement, open at the top, and with a small opening in front through which the fire is fed and which serves as a draft. The walls are about 6 inches thick. The outside dimensions of the fire box base are 3 feet 4 inches by 2 feet 6 inches. The inside dimensions of the fire box are 2 feet 3 inches by 1 foot $9\frac{1}{2}$ inches by 1 foot 4 inches. Across the top of the fire box there were laid, while the cement was still soft, some old grate bars from a small steam boiler, which had been discarded and thrown on the dump heap. These were set close together and held firmly in place when the cement hardened. They form the grate on which the material to be incinerated is thrown. These old boiler grate bars, besides costing nothing, had another advantage; namely that of their thickness and weight. When they become thoroughly heated from the fire below they will hold the heat for a considerable time, charring and burning the animal material above.

The incinerating chamber proper was made from galvanized iron by a local tinsmith. This consists of a rectangular box having the following dimensions: Length 2 feet 2 inches; width 1 foot 10 inches; height 1 foot 6

Fig. 5.—Photograph of crematory described in text. Note cement base, with opening in front into fire box: galvanized iron cremating box on top; cover of cremating box.

inches. In the top of this is cut a round hole, 12 inches in diameter which is protected by a hinged cover 15 inches by $14\frac{1}{2}$ inches. This galvanized box has no bottom. It is placed on top of the grate bars, and held firmly in place by cement worked up around its lower edges. At the back end of this iron cremating box is an opening for a stove pipe, which is necessary in order to give the proper draft. It is found in practice that only a short piece of stove pipe is necessary to get sufficient draft to make a very hot fire, which entirely consumes the birds in a few hours. The funnel may best be left removable so that when the crematory is not in use it can be taken off and stored inside the wooden box, which then sets over the galvanized iron portion to protect it from the weather.

It is important in locating a crematory of this kind to plan matters so that there will be good drainage from around it. In particular pains should be taken to insure that water does not run into the firebox and freeze during the winter.

In operation the apparatus works as follows: Dead birds are thrown into the incinerating chamber through the opening in the top and the lid closed, while a wood fire is burning in the fire box below. The aim should be to use dry wood and get a quick and very hot fire. This first roasts the material and then chars it, and finally reduces it to fine ashes.

VII. Isolation of Sickness

Whether one expects to treat the bird or to kill it, *every individual that shows signs of sickness should be removed from the general flock.* When the bird has been isolated a decision as to what will be done about the case can be reached at leisure, and in the meantime the flock is not subjected to the danger of infection. This is an important matter with young chickens as well as with adult stock.

The Essentials of Poultry Hygiene

To summarize this discussion of poultry hygiene and sanitation it may be said that the essentials in the hygienic and sanitary management of poultry are

1. Clean Houses.
2. Clean Air.
3. Clean Food.
4. Clean Water.
5. Clean Yards and Clean Range.
6. Clean Incubators and Brooders.
7. Clean Birds, Outside and Inside.

CHAPTER III

THE DIAGNOSIS OF THE DISEASES OF POULTRY

THE first thing that the poultry keeper whose birds are ill wants to know is: "*What ails my chickens?*" Before he can use this or any other book on poultry diseases effectively in getting advice for the treatment of disease he must diagnose the trouble. It is the purpose of this chapter to help him do this, and in this way make this book more useful to the practical poultryman. At the outstart it should be said that *the absolutely certain differential diagnosis of particular diseases of poultry, by the farmer or poultryman, either on the basis of external symptoms or post-mortem examination, is in nearly every case impossible. The best that can be done practically is to determine into what general class of diseases a particular trouble falls.*

There are two general sources of information upon which to base a diagnosis of disease. These are:

I. External symptoms.
II. Post-mortem examination.

EXTERNAL SYMPTOMS, WITH A TABLE TO AID IN THE IDENTIFICATION OF THE CHIEF CLASSES OF POULTRY DISEASES

There are certain external symptoms which are characteristic in a way of nearly all diseases. These symptoms merely indicate that the bird is *sick;* they are of *no value* for purposes of differential diagnosis.

The Diagnosis of the Diseases of Poultry

These general symptoms of illness may be described as follows: A sick fowl is usually quiet, and does not move about unless disturbed. It stands or sits with the neck contracted so that the head is pulled well in to the body, giving the bird a "humped up" appearance. The eyes are often closed, entirely or partly, giving the bird a sleepy appearance. Often the feathers are roughened and stick out all over the body. The comb and wattles may be dark or, on the other hand, may be very pale.

When a bird shows these general symptoms of illness it should be picked up and isolated and an effort made to obtain a more precise diagnosis. In doing this the following table of the chief external symptoms may be found of use.

This table aims to direct one to the discussion of *general classes* of disease. The identification of special individual diseases should be attempted only after reading over the chapters covering the general class involved. In general it should be kept in mind that *this table is not intended to tell the reader what the disease he finds is, but solely to tell him what parts of this book to read in any given case in order to make a diagnosis.*

TABLE OF EXTERNAL SYMPTOMS WHICH MAY BE OF SOME VALUE IN DIFFERENTIAL DIAGNOSIS

The numbers in parentheses denote the pages to be consulted.

Symptom	Diseases which the Symptom named may Indicate
Abdomen, swollen.	Peritonitis (77), Dropsy (80), White diarrhea (283).
Belching of gas.	Inflammation of crop (61).
Breathing abnormal (*i.e.*, too rapid, too slow, wheezing, whistling, snoring or in any way different from normal).	Diseases of the respiratory system (147), Crop bound (58), Arsenic poisoning (82), Pericarditis (182), Gapes (304), Air sac mite (180).

Symptom	Diseases which the Symptom named may Indicate
Choking.	Arsenic poisoning (82).
Comb, pale.	Tuberculosis (115), Dropsy (80), Air-sac mite (180), Infectious leukæmia (185), White diarrhea (283).
Comb, first dark then pale.	Roup (155).
Comb, first pale, but later dark.	Enteritis (67).
Comb, very dark.	Liver disease (87), Blackhead (94), Ptomaine poisoning (85), Congestion of lungs (177), Pneumonia (178).
Comb, yellow.	Liver diseases (87), Visceral gout (200).
Comb, with white, powdery scurf.	White comb (236).
Comb, with white spots.	Favus (233).
Constipation.	Simple constipation (69), Indigestion (70), Gastritis (63), Inflammation of Oviduct (262).
Convulsions.	Arsenic poisoning (82), Copper, lead or zinc poisoning (83), Epilepsy (196), "Harvest-bug" (227).
Cough.	Diseases of the respiratory system (147).
Crop, enlarged and hard.	Crop bound (58).
Crop, enlarged and soft.	Inflammation of crop (61), Enlarged crop (62), Inflated crop (63), Gastritis (63), Cholera (102).
Diarrhea.	Diseases of the alimentary tract (57), Arsenic poisoning (82), Copper, lead or zinc poisoning (83), Diseases of the liver (87), Blackhead (94), Tuberculosis (115), Cholera (102), Roup (155), White diarrhea (185), Coccidiosis (71), Mercury poisoning (83).
Droppings, blue.	Copper poisoning (83).

Symptom	Diseases which the Symptom named may Indicate
Droppings, bloody.	Diarrhea (64), Mercury poisoning (83), Blackhead (94), Enteritis (67), Arsenic poisoning (82), Ptomaine poisoning, (85) Diseases of the liver (87).
Droppings, bright emerald green.	Cholera (102), Copper poisoning (83).
Droppings, brownish followed by yellow diarrhea.	Diseases of the liver (87).
Droppings, hard and dry.	Constipation (69).
Droppings, mucus in.	Cholera (102), Diarrhea (64).
Droppings, sticky.	Simple diarrhea (64).
Droppings, slimy and yellow.	Nodular tæniasis (137).
Emaciation.	Tuberculosis (115), Aspergillosis (173), Visceral gout (200), Mites (213), White diarrhea (283).
Eye, expansion of pupil.	Arsenic poisoning (82).
Eye, sticky discharge from.	Catarrh (151), Roup (155).
Face, swollen.	Roup (155).
Fever, marked.	Peritonitis (77), Aspergillosis (173), Infectious leukæmia (185), Inflammation of oviduct, (262).
Head, warty nodules on.	Chicken pox (237).
Lameness.	Tuberculosis (115), Aspergillosis (173), Rheumatism (199), Scaly leg (216), Bumble foot (326).
Legs, roughened, with scales raised.	Scaly leg (216).
Mouth, mucous discharge from.	Congestion of the lungs (177), Pneumonia (178), Gapes (304).
Mouth, white cheesy patches in.	Roup (155), Canker (164).
Neck, bent backward.	Strychnine poisoning (84), Congestion of the brain (195), Wry neck (202), Pericarditis (182).
Neck, bent forward on breast.	Ptomaine poisoning (85).
Neck, limp.	Limberneck (199).
Nostrils, discharge from.	Diseases of the respiration system (147).

Symptom	Diseases which the Symptom named may Indicate
Paralysis.	Copper, lead or zinc poisoning (83), Strychnine poisoning (84), Apoplexy (194), Heat prostration (195), Polyneuritis (197).
Pulse, very rapid.	Hypertrophy of heart (184).
Saliva, copious secretion.	Arsenic poisoning (82).
Skin, puffed out in blisters.	Emphysema (304).
Skin, scaly and incrusted.	Body mange (226), Favus (233).
Staggering.	Congestion of the brain (195), Leg weakness (301), Ptomaine poisoning (85).
Thirst, excessive.	Hypertrophy of the liver (90), Peritonitis (77), Salt poisoning (81), Aspergillosis (173), Tapeworms (134), Cholera (102).
Tongue, hard and dry.	Pip (171), Diseases of the respiratory system (147).
Tumors, on head.	Roup (155), Chicken pox (237).
Urates, yellow.	Cholera (102).
Vent, mass of inflamed tissue projecting from.	Prolapse of oviduct (263).
Vent, skin inflamed.	Vent gleet (280).

POST-MORTEM EXAMINATION

Whenever a bird dies from a cause not entirely clear to the poultryman a post-mortem examination should be made in order to learn, if possible, from the condition of the internal organs what it was that caused death. The poultryman should familiarize himself with the appearance of the internal organs in a normal state of health, so that he may at once recognize any departure from these normal conditions.

The Normal Anatomy of the Domestic Fowl

Before undertaking a discussion of post-mortem appearance it is desirable to sketch in a brief way the most essential features of the normal skeletal and visceral anatomy of the fowl. If one will study this chapter with some care, and at the same time dissect a specimen, it will give him a sufficiently good understanding of the normal relations of the parts to enable him successfully to undertake for himself post-mortem examinations of his birds.

The Skeleton

The bones of birds (zoölogical class *Aves*) are in their structure somewhat different from the bones of other animals. The most essential difference consists in the fact that there are in the bones of birds, as a class, spaces which are normally filled with air, forming a part of the general air sac system connected with the lungs. The degree to which the bones have the capability of being filled with air varies considerably in the different orders of birds. In the genus *Gallus* (the domestic fowl) this possibility is small as compared with what obtains in flying birds, for example.

The skeleton of the cock is shown in Fig. 6. Its main divisions are: the *skull* (1); the *neck* (5); the *trunk* and the *limbs* (wings and legs).

Two parts are to be distinguished in the skull: first, the brain case or *cranium* proper (1) and, second, the *face*, including the *beak* (2, 3, and 4). As the skull bones grow together early, one usually does not see in the skull of the adult bird any division or sutures between the bones.

The trunk skeleton includes the backbone or vertebral column, the ribs, the breastbone, and the limb bones.

The skeleton of the *neck* (5) is made up of 13 separate

vertebræ of which the first, called the *atlas*, is the smallest. The vertebræ of the *back* (6) number seven, but they are usually nearly completely grown together into one single mass of bone. To these vertebræ of the back (called the *thoracic vertebræ*) are attached the seven pairs of *ribs* (7).

The *lumbar* vertebræ, of which only one or two remain, are in old birds grown together with the *sacral* vertebræ (17). Behind the sacral vertebræ come the *caudal* (18), which support the structure known, rather colloquially, as the "pope's nose." There are six *caudal* vertebræ.

The *pelvis* consists of three bones fused together: the *ilium* (19), the *ischium* (20), and the *pubis* (21).

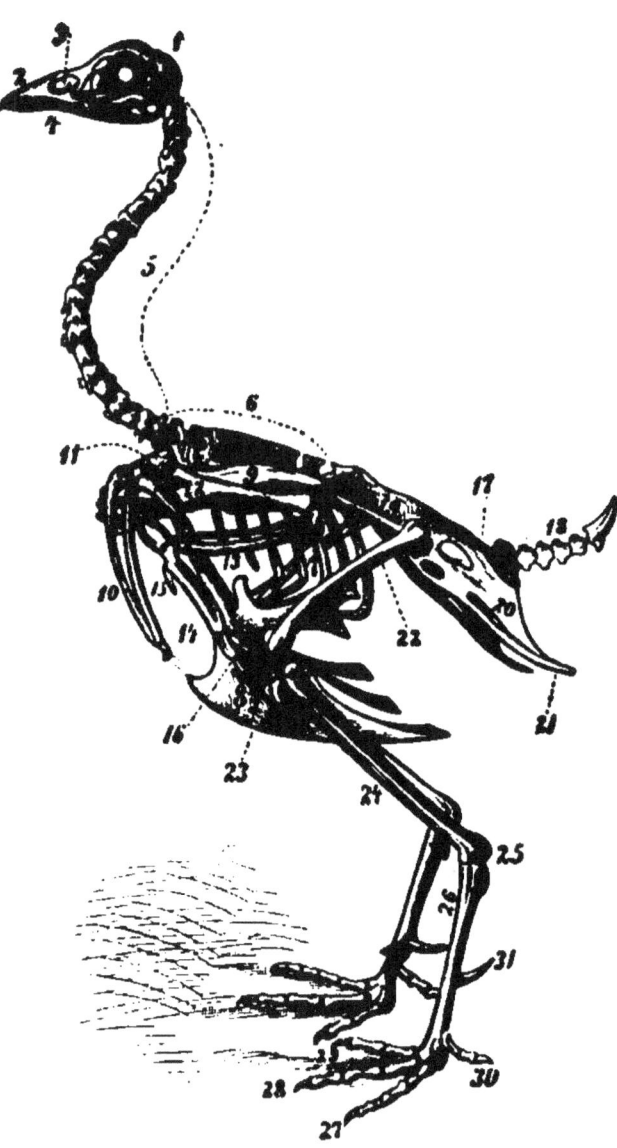

Fig. 6. — Skeleton of cock (*Gallus* sp.). For explanation of figures see text. (After Dürigen.)

The *breastbone* (8) bears a high and sharp bony plate, the *keel*, to which the breast muscles are attached. The wings are supported by the *pectoral* or *shoulder girdle* of bones. This girdle consists of the *shoulder blade* or *scapula* (9), the *coracoids* (11), and the "wishbone" or *furcula* (10). In young birds the furcula consists of separate paired bones, the *clavicles*, and the small median ossification, the *interclavicle*.

The *wings* include the upper arm, the forearm, the wrist, the hand, and the fingers. The *upper arm* bone, or *humerus* (12), is a single strong bone of the same length as the bones of the forearm. The *forearm* (13) contains two bones, the *radius* and the *ulna*. The wrist has only two bones not distinguishable in the figure. The *hand* (14) is made up from two bones which are united at both their upper and lower ends, but separated in the middle. In front of the larger of these two is the small one-jointed finger, the so-called *thumb* (15). The *second finger* (16), which is the longest and strongest, has two joints; the *third finger* is one-jointed.

The bones of the leg are homologous with those of the wing. The bone of the *thigh*, the *femur* (22), is a single bone. The *lower leg* (24), which in the normal position of the bird extends backwards and downwards from the *knee* (23), consists like the forearm of two bones: a large *tibia* and a very small splint-like bone, the *fibula*. At (25) is the *hock* or *ankle* joint. Below this come the bones of the foot. The first of these (26) is the *tarso-metarsus*. As an outgrowth from this bone is the *spur* (31). Of the four toes the inner or *hind toe* (30) has two joints; the *second* or inner front *toe* (29) has three joints; the *middle* front *toe* (28) has four; and the *outer* front *toe* (27) has five joints.

The Viscera

The main features of the normal visceral anatomy of the fowl are shown in Fig. 7, which represents a dissection from

44 *Diseases of Poultry*

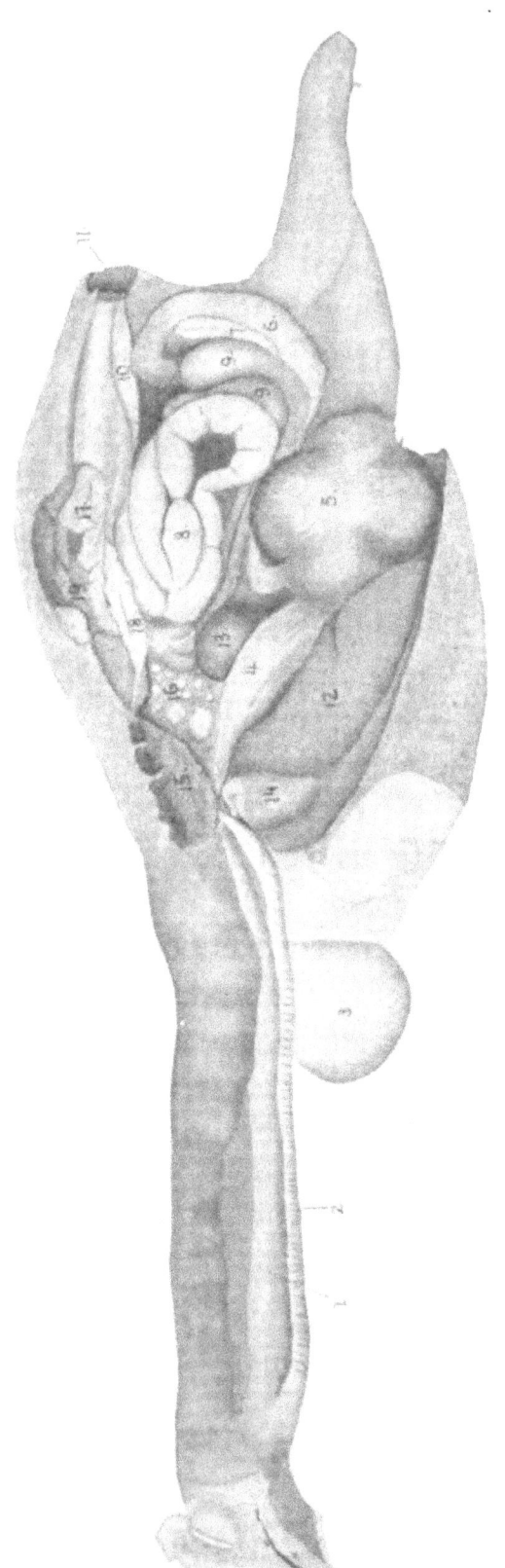

FIG. 7.—Dissection of hen, showing the normal visceral anatomy. For explanation see text. (Original.)

the left side of a hen carried out in such a way as to show the principal organs. Beginning at the anterior end we see, when the skin is removed from the bird, two tubes lying along the ventral aspect of the neck. These are the *esophagus* (1) and the windpipe or *trachea* (2). In the region where the neck joins the trunk there is a sac-like dilatation of the esophagus, the *crop* (3).

Proceeding backwards from the crop the alimentary tube passes through the thoracic cavity and as it enters the abdominal cavity widens out into the *proventriculus* (4), or glandular stomach. This opens directly into the *gizzard* (5) in which the mechanical breaking up and grinding of the food material is carried on.

After leaving the gizzard the food passes into the intestine. The first portion of the intestine, called the *duodenum* (6), forms a U-shaped loop. It consists of a left or proximal and a right or distal limb. Between these two limbs is situated an important digestive gland, the *pancreas* (7). Without any line of demarcation the duodenum is continued as the long, *small intestine* (8) back to the point of origin of two out-pocketings of the alimentary tract called the *ceca* (9). The point of origin of these organs marks the separation between the small intestine lying in front of them, and the large intestine or rectum (10) lying behind them. The rectum opens into the *cloaca* (11), a somewhat dilated chamber which opens externally by the vent. Into the cloaca open besides the rectum the ducts of the urinary and genital systems of organs.

On the ventral side of the abdominal cavity is seen the large *liver* (12), consisting of a large right lobe and a smaller left lobe. Attached by a fold of membrane to the proventriculus is seen the *spleen* (13), a small, ovoidal, red body. Just in front of the liver lies the *heart* (14), inclosed in a thin membranous sac, the *pericardium*. The *lungs* (15) are light red spongy bodies which may be seen by turning aside the heart, closely attached to the dorsal body wall in the thoracic region.

Just behind the lungs is seen on the left side of the body in the female the ovary (16). Except in rare cases, the ovary on the right side of the body in the domestic fowl degenerates during embryonic life and only the left ovary remains to produce eggs. It is made up of a number of spherical bodies of various sizes called ovarian eggs or *oöcytes*.

Behind the ovary is the *oviduct* (17), which in a laying hen is a much coiled tube with thick glandular walls. In this tube the albumen and other envelopes of the egg are manufactured. At the anterior end of the oviduct is the *infundibulum* (18) or funnel. (For further description of the reproductive organs see p. 245.)

Above and behind the ovary and oviduct lies the left kidney (19), a dark red body closely attached to the dorsal body wall. It and its fellow on the opposite side, the right kidney, consist of three well marked lobes, an anterior, a middle and a posterior. These lobes are embedded in the cavities which are made by the pelvis and sacrum. From each kidney a tube, the *ureter* (2), leads back to the cloaca and through this tube the waste products discharged by the kidney are carried to the outside of the body.

The above account covers the chief visceral organs. Any one wishing to go further into the study of the anatomy of the domestic fowl will do well to consult the larger handbooks of vertebrate comparative anatomy.

DIRECTIONS FOR MAKING A POST-MORTEM EXAMINATION

The poultryman about to undertake making autopsies on his birds should provide himself at the outstart with certain necessary articles. These should include as essential requisites, first, a good sharp knife with a pointed blade; second, a pair of bone forceps or else a pair of very heavy shears with short blades; third, a pair of medium size dissecting scissors, and, finally, a wooden table or dissecting board on which to work.

In making the post-mortem the dead bird should be laid with the breast uppermost on the table or dissecting board. The wings and legs should be spread out. A convenient way in which to hold the bird is to drive a small shingle nail through the tip of each wing and each foot, thus holding the body in the desired position. Then with a sharp knife an incision should be made through the skin in the general form of the letter U. The ends of the limbs of the U should be just behind the shoulder joints. The lower dip of the U should be just in front of the anus. It will not be necessary to remove the feathers in making this incision. They may be parted with the knife, which should be a heavy and sharp one. The incision should be carried through skin and body wall. In the region of the breast, where the muscles are thicker, the incision should be made clear through to the bone. The bone may then be divided in the same line by means either of the bone forceps or heavy shears already mentioned. An incision made as described will free the whole ventral body wall except at the anterior end. To free it at the anterior end it is necessary to use the bone forceps and cut straight across the anterior end of the body in front of the shoulders. Then the whole ventral body wall may be lifted off and the organs exposed in their natural position. In doing this one should be careful not to injure the heart or any other organs.

One may then proceed to the examination of the different organs for evidence of disease. Taking first the liver; one should note whether it has any spots or is discolored, or whether it is of a soft, friable consistency, a condition known as "punky" liver. The healthy liver should be a rich chocolate brown color, free of spots and discolorations. There may of course be post-mortem discolorations due to escape of bile or other causes, but these may be easily recognized as such. Spots on the liver may be of several sorts, either small or large nodules of whitish substance studded in the liver tissue, or simply areas of different color from the rest of the liver. Again there may be all over the surface of the liver a deposit of white material, which indicates excess urates from defective kidney function. As to consistency, the liver should be firm to the touch, and not easily friable or "punky."

The heart should be free of excessive fat and without tubercles. One of the most common diseased conditions of the heart is hypertrophy, wherein the heart is larger than normal. One should accustom himself to recognize the normal size of the heart and then in a post-mortem examination he can tell at a glance whether the heart is unduly enlarged.

The lungs should be examined for evidence of congestion, old or recent, and aspergillosis. Tuberculosis of the lungs is relatively rare in poultry. In the majority of attacks of tuberculosis the lungs are usually among the last organs to be affected.

Crop and gizzard are opened to determine whether anything in the recently ingested food has been causing trouble. It is also well to split open the trachea or windpipe to see whether any foreign bodies are present, or whether there is any accumulation of cheesy pus, indicating some form of roup, particularly the diphtheritic. Also, of course, a lookout should be kept for parasitic worms in opening any of the abdominal organs.

In the abdominal region, after the liver has been removed or turned forward out of the way, the spleen may then be seen. This is a small dark organ which is usually one of the first to be attacked in a case of tuberculosis. Where this disease is suspected, careful examination of the spleen should be made to see whether any small white nodules or tubercles are present.

The intestines should be examined for evidences of congestions, presence of parasites, or presence of tubercles.

After having been examined the intestines may be removed and thrown away. Then there will be exposed the urinary and reproductive organs. In many cases death in hens is due to some impairment of the egg producing organs. One should look first for tumors on the ovary, which may generally be distinguished as dark colored bodies attached to the ovary, looking very much like a yolk in process of formation which has gone bad. The oviduct should be examined for evidence of rupture of its walls, or the presence of concrements.

The kidneys have normally a dark red mottled appearance. In cases of disease the most marked change to be seen is one of color. The kidneys may thus be either extremely congested and much deeper color than the normal, or, on the other hand, they may be pale and take on a yellowish hue. Again they may be covered with a heavy white deposit of urates.

At this point one may make a thorough examination of the peritoneum or lining membrane of the body cavity to see whether or not this is diseased. In a condition of health it is a glistening, thin, transparent membrane which lines the whole of the body cavity and covers the viscera. In cases of disease or abnormal conditions which cause peritonitis, it becomes a thickened whitish or yellowish and opaque membrane.

If anything like roup or any other disease of the air passages is suspected, examination should be made of the nasal passages in the head. This can best be done by opening the lower jaw and then with a sharp knife making a slit straight back from the corners of the mouth so to fold the lower jaw back on the throat and expose the larynx and upper end of the trachea. At the same time by cutting away the roof of the mouth one can examine the nasal passages proper.

THE DIAGNOSTIC VALUE OF CERTAIN POST-MORTEM APPEARANCES

In the table which follows the attempt has been to do for the post-mortem diagnostic signs the same sort of thing as was done in an earlier section for the external signs of disease. There are listed in the table the more striking post-mortem appearances which the poultryman is likely to meet, together with an indication of the disease which may have been the cause of each appearance. It is hoped that this table may prove useful.

TABLE OF POST-MORTEM APPEARANCES WHICH MAY BE OF SOME VALUE IN DIFFERENTIAL DIAGNOSIS

The numbers in parentheses denote the pages to be consulted.

Post-mortem Appearance	Disease which the Appearance named may Indicate
Blood, clotted in vessels.	Thrombosis (185).
Brain, blood clot on.	Apoplexy (194).
Bronchi, filled with mucus.	Pneumonia (178).
Ceca, inflamed.	Coccidiosis (285), Blackhead (94).
Ceca, partially filled with grayish soft material, *not* cheesy (chicks).	Bacillary white diarrhea (287).
Ceca, thickened and distended with pasty or cheesy mass.	Coccidiosis (285), Blackhead (94).
Gall bladder, distended.	Jaundice (94).
Heart, exudate in pericardial cavity.	Fowl plague (112), Pericarditis (182).
Heart, lining membrane reddened.	Endocarditis (183), Myocarditis (183).
Heart, punctiform hemorrhages of.	Cholera (102), Fowl plague (112).
Intestine, bearing nodules.	Tuberculosis (115), Tumors (312).
Intestines, congested.	Ptomaine poisoning (85).
Intestines, filled with mucus.	Enteritis (67).
Intestines, inflamed.	Enteritis (67), Mercury poisoning (83), Blackhead (94).
Intestines, mucosa bleeding.	Cholera (102).
Kidneys, bearing cheesy nodules.	Tuberculosis (115).
Kidneys, congested.	Ptomaine poisoning (85).
Kidneys, enlarged.	Diseases of the kidneys (199).
Kidneys, filled with whitish crystalline deposit.	Mercury poisoning (83), Diseases of the kidneys (199).
Kidneys, inflamed.	Cholera (102).
Liver, chalky deposit on.	Gout (199).

Post-mortem Appearance	Diseases which the Appearance named may Indicate
Liver, congested.	Ptomaine poisoning (85), Diseases of the liver (87), Cholera (102), Infectious leukæmia (185).
Liver, enlarged.	Enteritis (67), Diseases of the liver (87), Infectious leukæmia (185).
Liver, greasy.	Hypertrophy of liver (90).
Liver, pale with streaks of red (chicks).	Bacillary white diarrhea (287).
Liver, shrunken.	Enteritis (67), Diseases of the liver (87).
Liver, spotted (or marbled).	Coccidiosis (285), Diseases of the liver (87), Blackhead (94), Aspergillosis (173), Cholera (102), Infectious leukæmia (185).
Liver, with raised nodules.	Tuberculosis (115).
Lungs, congested.	Pneumonia (178), Congestion of lungs (177), Cholera (102).
Lungs, dark color and solid.	Pneumonia (178).
Lungs, with cheesy nodules.	Tuberculosis (115), Aspergillosis (173).
Lungs, yellow membranous patches in.	Aspergillosis (173).
Ovary, with discolored tumorlike yolks.	Bacillary white diarrhea (287).
Ovary, without yolks.	Atrophy of ovary (251).
Oviduct, inflamed.	Diseases of the reproductive organs (245).
Peritoneum, covered with chalky deposit.	Gout (199).
Peritoneum, covered with opaque exudate.	Peritonitis (77).
Spleen, enlarged.	Enteritis (67).
Spleen, spotted.	Enteritis (67), Tuberculosis (115).
Ureters, yellow and distended.	Cholera (102).
Windpipe, yellow dust in.	Air-sac mite (180).
Windpipe, worms in.	Gapes (304).
Windpipe, yellow patches in.	Aspergillosis (173).

CHAPTER IV

POULTRY MATERIA MEDICA

IT is the purpose of this chapter to give an account of the drugs and remedies which the poultryman will find it well to be supplied with; directions for making various solutions; tables of weights and measures and the like.

THE MEDICINE CHEST

The following drugs and medicines will be found useful to have at hand.

Calomel (Sub-chloride of mercury). — This drug is chiefly useful for its effect on the liver. The dosage is anything up to 1 grain at a time. A dose of calomel should be followed in the course of two hours or so with a dose of castor oil.

Cayenne Pepper. — This is an excellent digestive and liver stimulant when given in the food in small quantities. It is also useful in cases of colds. In this case the pepper should be put in small gelatine capsules (size No. 4) which may be obtained from any druggist and a filled capsule then pushed far enough down the esophagus with the finger so that the bird will swallow it.

Catechu. — Bradshaw [1] says that this "in powder or tincture form in combination with powdered chalk is a good remedy for diarrhea. The average dose of powdered catechu is from 2 to 5 grains and of the tincture 2 to 5 drops."

[1] Bradshaw, G., "Poultry Farming." Department of Agriculture, N. S. Wales, Farmers' Bul. No. 51, p. 28, 1911.

Castor Oil. — Castor oil is used as a remedy for diarrhea and as an intestinal antiseptic. It may also be used in cases of crop-bound fowls, although for this purpose cotton seed oil will be found to be quite as satisfactory and very much cheaper.

Epsom Salts (Magnesium sulphate). — This is on the whole the most useful poultry yard drug. It is indicated in practically all cases of digestive disturbance and colds, bowel trouble, etc. The standard dose for an adult fowl is from ½ to 1 teaspoonful.

The following table of doses of Epsom salts for young birds has been worked out by Gage and Opperman [1]:

Age of Bird	Amount per Bird in Grains	How Administered
1 to 5 weeks	10 grains	In feed
5 to 10 weeks	15 grains	In feed
10 to 15 weeks	20 grains	In feed
15 weeks to 6 months	30 grains	Two teaspoonfuls of water to every 30, 40 or 50 grains of salt.
6 months to 1 year	35 grains	
1 year and over	40–50 grains	

There are several ways of administering Epsom salts. It may be mixed with the drinking water, or a solution may be made with warm water and put down the throat of the bird. Probably, however, the best way to administer a dose to a large flock is to give the birds no food whatever on the day that they are to be given the Epsom salts until late in the afternoon. Then having determined the amount of salts to be used for the whole pen of birds at the rate of from ½ to 1 teaspoonful per bird, dissolve this amount

[1] Gage, G. E., and Opperman, C. L., "A Tapeworm Disease of Fowls." Maryland Agr. Expt. Stat. Bulletin 139, pp. 73–85, 1909.

in water and use this solution to mix up a wet mash. Any ordinary dry mash mixture of bran, meal and other ground grain may be used for the purpose. The wet mash so prepared should be divided into several lots and put in different places in the pen so that all the birds will get a chance at it. This method insures a more even dosage through the flock than any other we have tried.

Cotton-seed Oil. — A bland oil like cotton-seed oil (salad oil) is useful in many ways about the poultry yard. In treating prolapse of the oviduct, crop-bound condition, and in other cases, the oil may be used to good effect as a simple lubricant. Bradshaw says that in the case of eye trouble it may take the place of a simple lotion.

Bichloride of Mercury, 1 to 1000 Solution. — To make this the simplest way is to buy of the druggist bichloride of mercury tablets, and ask him to label the box to show how much water a tablet must be dissolved in to make a 1 to 1000 solution. If one desires to mix it up for himself ask the druggist to make up some *1 gram* ($15\frac{1}{2}$ *grain*) powders of bichloride of mercury. Dissolve 1 of these powders in a quart of water. Put in enough laundry bluing so that the color will be deep blue. Then the solution, which is highly poisonous, will never be mistaken for water.

"1 to 1000 bichloride" is a germicide and disinfectant for external use, cleansing wounds and the like.

Medicines in Tablet Form. — One of the most convenient forms in which medicines may be administered is in tablets. Wholesale and mail-order drug houses carry extensive lines of these graded as to dosage. They may be administered to poultry very easily and conveniently by holding the bird's mouth open with one hand and with the other thrusting the tablet far enough back in the throat so it will be swallowed.

The following list of tablets will be found useful to the poultryman. They fairly well cover the medicines recommended in the body of this book. Any poultryman may get these either from his local druggist, or if he cannot furnish them, they can be purchased by mail at approximately the prices named.

	APPROXIMATE PRICE PER 1000
Sodium salicylate, 3 gr.	$ .70
(For use in rheumatism.)	
Aconite root, 1–10 gr.	.50
(For use in fevers.)	
Antiseptic tablets, *Blue*, Corrosive sublimate, 7.3 gr.; Ammonium chloride, 7.7 gr.	2.50
(For making 1 to 1000 bichloride solution. One tablet dissolved in 1 pint of water gives a solution of that strength.)	
Bismuth subnitrate, 1 gr.	.80
(For intestinal irritation.)	
Calomel, ¼ gr.	.40
Iron, Quinine and Strychnine.	.80
(For use as a tonic, dose 3 per day.)	

In administering tablets in the manner suggested care should be taken to see that they are swallowed, and not coughed up.

An Antiseptic Ointment for Use on Cuts and Wounds of All Kinds

The following ointment may be made up by the poultryman and will be found useful in the treatment of cuts, sores and wounds of all kinds of poultry and stock in general.

Oil of origanum	1 oz.
Cresol	¾ oz.

Pine tar............................ 1 oz.
Resin.............................. 1 oz.
Clean axle grease.................. 8 oz.

Melt the axle grease and resin and stir in the other ingredients. Pour off in a tin box or can to cool. In making this, clean axle grease from a freshly opened can should be used.

TABLES OF APOTHECARIES' WEIGHTS AND MEASURES AND THEIR METRIC EQUIVALENTS

APOTHECARIES' WEIGHTS

POUND	OUNCES (TROY)	DRAMS	SCRUPLES	GRAINS	GRAMS
1 =	12 =	96 =	288 =	5760 =	373.23
	1 =	8 =	24 =	480 =	31.10
		1 =	3 =	60 =	3.9
			1 =	20 =	1.30

APOTHECARIES' MEASURE

GALLON	PINTS	FLUIDOUNCES	FLUIDRAMS	MINIMS	CUB. CM.
1 =	8 =	128 =	1024 =	61440 =	3785.00
	1 =	16 =	128 =	7680 =	473.11
		1 =	8 =	480 =	29.57
			1 =	60 =	3.75

COMMON MEASURE

A *teacup* is estimated to hold about 4 fluidounces, one gill.
A *wineglass* is estimated to hold about 2 fluidounces.
A *tablespoon* is estimated to hold about ½ fluidounce.
A *teaspoon* is estimated to hold about 1 fluidram.

CHAPTER V

DISEASES OF THE ALIMENTARY TRACT

THE arrangement of the digestive organs in birds differs from that in other domestic animals in that the mastication of the food does not take place in the mouth. The food of birds, consisting mainly of grains and seeds, is swallowed whole into the crop. It remains here until it is completely softened by the juices secreted by this organ. The food then passes into the stomach (proventriculus), where it is mixed with still other juices, and then into the gizzard. The muscular walls of the gizzard grind the softened food against the small pebbles (grit) which the bird picks up, until it becomes a paste. This paste is then passed into the intestines and mixed with the secretions from the liver, pancreas and the intestines themselves. The nutritive elements of the food are transferred through the intestinal walls, by means of the activity of the cells composing these walls, into the blood and are carried to various parts of the animal to be used in building up the tissues.

In the wild state birds are forced to hunt for their own food. They go about gathering in a few seeds here and there but probably at no time is the crop overloaded. Under conditions of domestication birds are fed only once or twice a day and thus the crop is often gorged with a day's supply of food. Further the lack of sufficient grit, lack of exercise and the feeding of rich, soft mashes cause the birds to be predisposed towards indigestion. Under these conditions poultry are subject to a large number of disorders of the digestive system.

DISEASES OF THE CROP

Impacted Crop (Crop Bound)

In general two immediate causes may be given for birds becoming crop bound. (1) The thin muscular walls may be paralyzed either through over-distention with dry grain or through some disease, as cholera and diphtheria. (2) The opening into the lower portion of the esophagus may become clogged by long straws, feathers or other substances. In either case the crop fails to empty itself while the bird continues to eat until the crop is greatly distended and packed solid.

Impacted crop is a common disease of poultry. A large number of things have been assigned as a cause for this trouble. It is probable that the real cause lies in low vitality due to improper feeding and indigestion. Every poultryman knows that very often fowls will eat large quantities of hay, straw, strings, feathers, etc., without showing the least inconvenience. If the digestive organs are in the proper health and tone they will usually take care of any overloading of the crop. It is only when the tone of the digestive system has been lowered by improper feeding, housing or by some disease that the crop fails to perform its usual function. Occasionally a case of impaction may properly be attributed to overloading the crop with indigestible matter. Such cases will occur only rarely and sporadically. If many crop bound birds appear in a flock it may be taken as certain that something more fundamental is the cause.

Symptoms. — The first symptom is a loss of appetite or an effort of the bird to swallow without being able to do so.

The crop is seen to be very large and much distended with contents which are more or less firmly packed together. If permitted to continue, the condition becomes aggravated, the breathing difficult, and death may result.

Treatment. — If a large number of crop bound birds occur in a flock, it should be taken as a sign that something is wrong in the management. Measures should be taken to correct errors in feeding and thus give the birds a more vigorous digestion. In such epidemics other evidences of indigestion are usually present and the particular treatment of the flock will depend largely on these other symptoms. In general the birds should not be fed too much at any one time and they should be encouraged to take as much exercise as possible, and should have plenty of green food.

When a crop bound bird is found it must be treated individually. Treatment in such individual cases is quite often successful. The profitableness of such treatment must be decided by every poultryman for himself. If the crop bound condition is discovered and treated at the *beginning* of the trouble the bird will usually recover quickly and may make a profitable fowl. On the other hand if the condition has become chronic the vitality of the bird is greatly lowered. In this latter case it may recover but it will be a long time before it will repay the owner for his trouble and feed.

If swelled grain is the cause of the impaction the bird may often be successfully treated without an operation. In this case first give the bird a tablespoonful of castor oil. After allowing this a little time to work into the crop begin to knead the hard mass. After this mass has been softened hold the bird with head downward and attempt to work the grain out through the mouth. If unsuccessful in this or if the impaction is due to clogging with straw or other material it will be necessary to open the crop.

The operation for impacted crop is comparatively simple.

It will be easier if some person can hold the bird while another performs the operation. If assistance is not at hand the bird may be tied, back down, to a board or table. The operation should be done in a place as free as possible from dust and dirt. First, pluck out a few feathers in the median line of the crop. The feathers around the edge of the field of operation may be dampened to keep dust from them out of the wound. With a sharp, clean knife cut through the skin over the middle of the crop. This cut should be about 1 inch long. Then make an incision about $\frac{3}{4}$ of an inch long through the wall of the crop. The distention of the crop will cause the opening to gape, and the mass will be in plain sight. With a buttonhook, blunt pointed scissors, tweezers, or similar tools, take out the contents of the crop. This done, run the finger into the crop and make sure that there is nothing remaining to obstruct the outlet of the organ. After this is done thoroughly wash out the empty crop with clean warm (108° to 110° F.) water. The opening in the wall of the crop should be closed with 3 or 4 stitches, making each stitch by itself and tying a knot that will not slip. Then do the same thing to the cut in the skin. For stitches use white silk or (if nothing better can be obtained) common cotton thread, number 60.

The above operation is not a difficult one and is usually successful. Care should be exercised to have the hands and instruments thoroughly clean. In sewing up the wound care should be taken that dirt, ends of feathers, etc., are not drawn into the wound. Chickens are quite resistant to infection with ordinary bacteria, but the results will be uniformly better if care is taken to exclude all chance for infection. The edges of the skin should be well greased with vaseline. For the first day or two it is well to feed the bird only milk or raw eggs beaten together.

Inflammation of the Crop

Inflammation or catarrh of the crop usually accompanies more or less general disturbances of the digestive system. As a result of the irritated condition of the mucous membrane the functions of the crop are disturbed or arrested. This trouble, when not due to a generally run-down condition and lack of tone, is usually caused by eating moldy or putrid food and, especially, irritating mineral poisons. Unslaked lime, paint skins, and common salt are some of the more frequent causes. Worms in the crop may also cause an inflammatory condition. It also occurs as a complication with diphtheria, cholera, etc. Inflammation of the crop is usually accompanied by more or less severe inflammation of the other regions of the digestive tract. The cause which irritates the crop also disturbs the mucous linings of the other regions.

Diagnosis. — The most prominent symptom is distention of the crop, and on examination the swelling is found to be soft and due to accumulated liquid or gas, mixed with more or less food. The birds are dull, indisposed to move, and there is belching of gas, loss of appetite and weakness. Pressure upon the crop causes the expulsion through the mouth of liquid and gas having an offensive odor, due to fermentation.

Treatment. — The first step in the treatment of this disease is to empty the crop as completely as possible. This can be done by holding the bird head downward and carefully pressing and kneading the crop. After most of the contents have been expelled in this way give the bird several spoonfuls of lukewarm water and then empty the crop as before. Give a slight purgative such as a small teaspoonful of castor oil. The bird should be kept without food for 12 to 20 hours and then fed sparingly on soft, easily digested material. Two grains of subnitrate of bismuth and $\frac{1}{2}$ grain of bicarbonate of soda in a teaspoonful of water will relieve the irritation and correct the acidity. Salicylic acid, 1 grain to an ounce of water, is also recommended. The dose is 2 to 3 teaspoonfuls. The feeding of mucilaginous fluids such as barley-water, thin solution of gum, etc., is recommended. If the inflammation is due to eating poisons antidotes as given farther on (Chapter VI) should be used.

If inflammation of the crop is at all general throughout the flock an effort should be made to remove the cause. It is well to change the feed and give the birds more exercise. The addition of fine charcoal (small chick size) to the mash will often be of service, as the birds eat more of it in this way than when the charcoal is in a box by itself.

Enlarged Crop

One sometimes finds a bird with a very much enlarged, pendulous crop. This loose baggy condition is usually permanent, but in the majority of cases it does not cause the bird any serious inconvenience.

The cause of this enlarged or slack crop is usually said to be overfeeding at irregular intervals. It is probable that overloading of the crop alone is not the only cause. Overloading accompanied by indigestion or some general disturbance of the digestive organs may result in a sort of paralysis of the crop muscles. It is not improbable that many birds showing enlarged crop have suffered with a mild case of impaction and have finally recovered without assistance.

An enlarged crop and an enlarged "baggy" abdomen are frequently associated in the same bird. These are usually said to be due to too heavy feeding without sufficient intervals between meals and without sufficient exercise.

Treatment. — As stated above, a "baggy" crop often gives little or no apparent inconvenience to the fowl. In the case of a very valuable bird it might be worth while to operate. It is said that this defect can be remedied by cutting out of the enlarged portion of the crop a diamond or oval shaped piece of tissue about 2 inches long and 1 inch wide. The edges should be sewed together and treated as directed for impacted crop (cf. p. 60). The general surgical methods described in the chapter on Poultry Surgery (Chapter XXI) should be followed.

Inflated Crop

Occasionally birds both old and young are found with enormously inflated crops. This condition is due to the pressure of gas forming bacteria. It is probably caused by eating decayed food. The remedy for this trouble is first to remove the cause and then give a mild intestinal antiseptic in the drinking water, such as 1 to 10,000 bichloride of mercury or 1 to 500 carbolic acid.

DISEASES OF THE STOMACH (PROVENTRICULUS)

Inflammation of the Stomach — Gastritis

The stomach or proventriculus in fowls is a rather small organ. It is a thick, glandular walled section of the alimentary canal lying between the crop and the gizzard. Inflammation of this organ is usually associated with a similar disturbance of the crop. In a few cases there appears to be inflammation of the stomach alone. Diagnosis in this case is very difficult.

The cause of gastritis is usually regarded as the same as that of inflammation of the crop (cf. p. 61).

Diagnosis. — In general the symptoms are very similar to those in cases of inflammation of the crop (see p. 61). The birds present the general appearance of being sick, viz., loss of appetite, indisposition to move and roughness of plumage. Constipation quite often accompanies gastritis. However, if the inflammation extends to the intestines there may be diarrhea.

Treatment. — The most important thing in the treatment of this kind of a disease is to ascertain and remove the cause. Medical treatment without removal of the cause will do but little good. The kind of food which the birds have access to should be examined and any changes made which might remove the cause of the trouble. The addition of fine (chick size) charcoal to the mash and the generous use of good green food are recommended. For a time the birds should be fed often, giving only a small quantity at a time. A good cooked food is often more easily digested and will aid in stopping the irritation. Give the birds barley water or milk to drink, or add 20 grains of bicarbonate of soda (baking soda) to a quart of drinking water. Rice water to which $\frac{1}{4}$ grain of arsenite of copper to each quart has been added is also recommended. In severe cases give 2 grains of subnitrate of bismuth 3 times a day in a teaspoonful of water. Counteract constipation with Epsom salts (20 grains) or castor oil (one teaspoonful) once a day as long as may be necessary.

DISEASES OF THE INTESTINES

Simple Diarrhea

In many fowls a condition of mild diarrhea is chronic throughout the lifetime of the bird. Again birds often acquire a slight diarrhea which will last for a longer or shorter time, but never becoming severe. In either of these cases the bird shows no symptoms of disease other than the watery droppings. No doubt such attacks are in some degree detrimental to the best health of the bird. In most cases of this simple diarrhea the bird will recover without any treatment. Nevertheless the careful poultryman will watch his dropping boards for signs of "looseness." When such are found in any quantity the methods of feeding and housing should be carefully examined to see if the cause does not lie in them.

The normal droppings of a fowl are almost dry and retain the shape in which they are voided. They are easily removed from the dropping board and leave little or no stain. About $\frac{1}{3}$ of the normal dropping consists of a whitish substance. This is the uric acid and urates excreted by the kidneys and removed from the cloaca along with the feces.

One not infrequently finds droppings which are more watery than the normal. These have a tendency to stain the dropping boards and do not retain the shape in which they are voided. This condition is best described as "looseness" and is quite different from true diarrhea. Looseness is not accompanied by offensive odors. Looseness of the bowels may be caused by a large number of things, such as a slight change of food, an additional amount or a new kind of green food, etc. Some individual birds appear to void loose, slightly watery droppings throughout life. Such individual differences are not uncommon among other animals.

Looseness of the bowels is a condition which need cause no alarm, but when droppings are found which are sticky or liquid in consistency and have a yellow brown or greenish color accompanied by an offensive odor it is time to look after the cause. The evacuations described above indicate some form of true diarrhea. Very often the watery evacuations contain mucus and in the more severe cases small clots of blood. A true diarrhea is nearly always shown by the soiling of the bird's feathers.

Diarrhea may result simply from an upsetting of the digestive organs due to improper feeding or it may be a symptom of some more serious disease. Simple diarrhea may arise from the presence of indigestible matter in the alimentary canal, it may be due to exposure to heavy rains or to drafts in the roosting house. In the latter cases a cold develops which affects the bowels rather than the head and lungs. Diarrhea from colds occurs much more frequently than is generally supposed. This form of diarrhea can often be recognized by the greater amount of frothy mucus in the excrement. Young stock are much more susceptible to diarrhea from colds than are adult birds.

Among other common causes of simple diarrhea may be mentioned soured or decomposing food, too much green food at irregular times, too free use of animal food, allowing the birds access to water which has become soiled with excrement and allowed to stand in the hot sun until about putrid. Whatever may be the inducing factor the immediate cause is excessive bacterial fermentation in the alimentary canal.

Treatment. — Simple diarrhea will usually require no treatment other than removing the original cause. This latter is by far the more important thing to be done. If neglected, the condition may become chronic and may result in more serious disturbances of the alimentary system. It is often beneficial to replace part of the bran in the mash with middlings or low grade flour. Where in addition medical treatment seems desirable the first thing to do is to remove the fermenting material from the intestinal canal. This can be done with Epsom salts, using a small half teaspoonful to each bird. This should be dissolved in water and used to mix the mash. If more convenient, a teaspoonful of castor oil may be given each bird. If the diarrhea is persistent, 3 to 6 drops of chlorodyne is said to be an unfailing cure.

Enteritis — Dysentery

For practical purposes we may associate most of the severer forms of diarrhea with the above names. Simple diarrhea was defined as either a temporary or chronic affection of the intestines from which the bird appeared to suffer but little. Practically its only symptom is the watery or discolored discharge. Under the names of enteritis, dysentery or severe diarrhea there are listed several of the more serious infections of the intestines. From the medical standpoint enteritis is the name given to affections of the small intestines, while dysentery is applied to the disease in the large intestine. The latter is usually accompanied by mucous and bloody discharges. In the diseases of poultry, however, it is hardly necessary for any one other than a pathologist to distinguish between these different forms.

Etiology. — A variety of causes are responsible for these more acute forms of intestinal trouble. It may be a bacterial infection coming from filthy conditions. Foul drinking water, putrid meat or decaying food of any sort may be predisposing causes. Toxic enteritis or poisoning is caused by the birds eating such things as paint skins, lye, unslaked lime, salt, ergot of rye, arsenic and copper (in spraying mixtures) (cf. p. 81). Further simple diarrhea may develop into the more acute form. This latter is due to improper food, water or housing, and is probably closely associated with bacterial enteritis. Various intestinal parasites may cause severe diarrhea.

Diagnosis. — It is often very difficult to distinguish between the different infections of the intestines in the living birds. In all these cases the birds are inactive and appear sleepy. The comb is often pale and bleached in the earlier stages but becomes dark purplish red later. Usually the birds will not eat, but occasionally they show an abnormal appetite. There is always a marked diarrhea which may vary in color from whitish to greenish brown or red. In the more severe cases blood clots are found. These differences in the appearance of the discharges indicate to some extent which portions of the alimentary tract are involved in the disease. In the majority of cases the birds will be sick for many days or weeks before death takes place. Post-mortem examination shows usually an enlarged liver and spleen. If the bird has been sick for a long time the liver may appear shrunken. The intestines are full of mucus and inflamed.

Treatment. — If possible the cause of the trouble should be ascertained and removed. This is by all means the first and most important step to take. It is useless to spend valuable time in doctoring sick birds while the conditions which gave rise to the trouble are still present. In bacterial enteritis sick birds should be removed from the flock as soon as noticed. Houses and runs should be cleaned up and disinfected. Drinking vessels and food troughs should be scalded daily. Potassium permanganate should be used in the drinking water (cf. p. 25). Mix powdered charcoal with the mash. Feed less bran and more middlings in the mash. Do not feed too heavily.

After attending to the above hygienic measures the birds should be given a good physic. A teaspoonful of Epsom salts to each fowl, dissolved in water and mixed in the mash, is the most convenient way of treating a large number of birds.

For medical treatment the following may be recommended: Subnitrate of bismuth, 3 grains; powdered cinnamon or cloves, 1 grain; powdered willow charcoal, 3 grains. Give twice a day mixed with food or made into pills with flour and water.

Subnitrate of bismuth, 3 grains; bicarbonate of soda, 1 grain; powdered cinchona bark, 2 grains; mix and give 3 times a day in a paste made with wheat flour. When diarrhea is arrested, bismuth and soda are no longer needed.

It is often worth while to give a good tonic or condition powder to aid the birds in getting their digestive organs in order again. The following tonic is recommended by Salmon: Powdered fennel, anise, coriander, and cinchona — each 30 grains; powdered gentian and ginger each 1 dram, powdered sulphate of iron, 15 grains. Mix and give in the feed so that each fowl will get 2 to 14 grains twice a day. (For another tonic see p. 71.)

Constipation

Constipation occurs in adult fowls far less often than diarrhea. It frequently passes unnoticed unless very severe. This trouble is much more common in young stock than in grown birds. In adult fowls it often occurs in connection with indigestion, gastritis, or peritonitis. Among the specific causes of constipation lack of exercise and lack of green food are probably the most important. Occasionally intestinal worms will accumulate until they block the intestine. Sometimes following a diarrhea the vent will become obstructed with dried evacuation. This is particularly apt to occur in young birds which do not roost. It is one of the symptoms of white diarrhea.

The *symptoms* of constipation are painful and ineffective efforts to evacuate the bowels. In the worst cases the vent becomes completely plugged with dry, hard feces. The birds appear dull, listless and without appetite.

Treatment. — When the vent is plugged with dried feces

the first thing to be done is to remove this. This can usually be done by soaking the mass with warm soapsuds. As soon as this is loosened a little the feathers can be clipped and the entire mass removed. If the case is of long standing the cloaca may also be filled with hard excrement. This can sometimes be softened by injecting warm soapsuds or a little olive or sweet oil. In all cases a purgative should be given such as castor oil, Epsom salts or calomel.

Indigestion

Birds frequently suffer from disorders of the digestive system which are not easily classified under any of the diseases so far treated. Simple indigestion or dyspepsia most frequently results from overfeeding, and the feeding of ground grains and meat without sufficient green food are some of the causes usually given.

Symptoms. — The birds are dull and listless. They are inclined to sit on the roosts, and usually have but little appetite. Occasionally birds suffering from indigestion have an abnormal appetite and will eat ravenously quantities of foods which furnish but little nourishment, *e.g.*, grit. Indigestion is often accompanied by either constipation or diarrhea. In the latter case the symptoms are similar to those described under simple diarrhea (p. 64).

Treatment. — In treating indigestion it is important to observe the general rules of hygiene (cf. Chapter II). The house should be clean and as free from dust as possible. Sunshine should be able to reach every corner of the pens. The water dishes should be kept thoroughly clean and the supply of water should be kept pure and fresh. Use potassium permanganate in the drinking water as directed on p. 25. Use well balanced rations and feed at regular hours. Put fine (chick size) charcoal into the mash in considerable quantity. Enough should be used to make the mash decidedly black. This is a very important measure for the treatment of indigestion. Give the birds plenty of exercise. A small amount of a good stock tonic may help to bring the birds back into proper vigor. The following formula has frequently been used with good success.

Pulverized Gentian.....................1 lb.
Pulverized Ginger......................¼ lb.
Pulverized Saltpeter...................¼ lb.
Pulverized Iron Sulphate..............½ lb.

These substances can be procured from any drug store and mixed by the poultryman. Use 2 to 3 tablespoonfuls of the tonic to 10 quarts of dry mash.

Recovery from indigestion may also be hastened by the following treatment. For the first week after the trouble has been discovered add one teaspoonful of Epsom salts to each quart of drinking water. Follow this for two weeks with ⅛ grain of strychnine to each quart of drinking water.

Coccidiosis

This disease is produced by small protozoan parasites which attack various regions of the intestinal tract. There are a large number of different species of "coccidia" which frequently attack birds and the smaller mammals, such as rabbits, rats, and mice. They are very destructive to young birds and are said by some investigators to be the cause of one form of white diarrhea in young chickens (see Chapter XVIII).

Many different species of birds are attacked by coccidiosis. Pigeons are particularly liable to the disease, and are frequently responsible for the outbreaks in the poultry yards. The transmission of the contagion from diseased to healthy birds occurs by contamination of the food, water, gravel, and other substances taken into the digestive organs. The coccidia multiply with great rapidity in the intestines of diseased birds, and enormous numbers are discharged with the droppings and are carried on the birds' feet to the feed troughs and drinking fountains unless these are well protected and of such form that they cannot be reached by the feet. Under any circumstances they are spread over the floor of the houses and the surface of the runs, and many will be picked up with gravel, grain, and other substances. The germs are found in the part of the small intestine nearest to the gizzard, where they cause inflammation, with redness and thickening of the intestinal wall. They are also found in the ceca, which are frequently thickened and distended with a whitish, yellowish, or greenish yellow, pasty mass. After two or three weeks the disease may extend to the liver and lungs, where it is recognized by whitish or yellowish spots or by large cheesy nodules. Geese are attacked by another species, which causes nodules in the kidneys.

The life history of a coccidium is very complicated, yet in order to combat this parasite most successfully it is necessary to know something of its life history. Figure 8 represents the different stages in the life history of one of these parasites. If one should examine with a microscope the contents of one of the ceca of a bird which died with a form of coccidiosis he would find forms somewhat like No. 1 in the figure.

These are the oöcysts or permanent cysts of the coccidium. The membrane around the outside of this cyst is very tough and will withstand almost all methods of disinfection. It will live and even grow in sulphuric acid. It can be killed, however, by drying. The size of one of these cysts is between $\frac{14}{25000}$ and $\frac{21}{25000}$ inch. If this cyst is placed under the right conditions for development the first step is for the protoplasm to divide into four spherical bodies which are called sporoblasts (Fig. 8, 2). Each of these sporoblasts

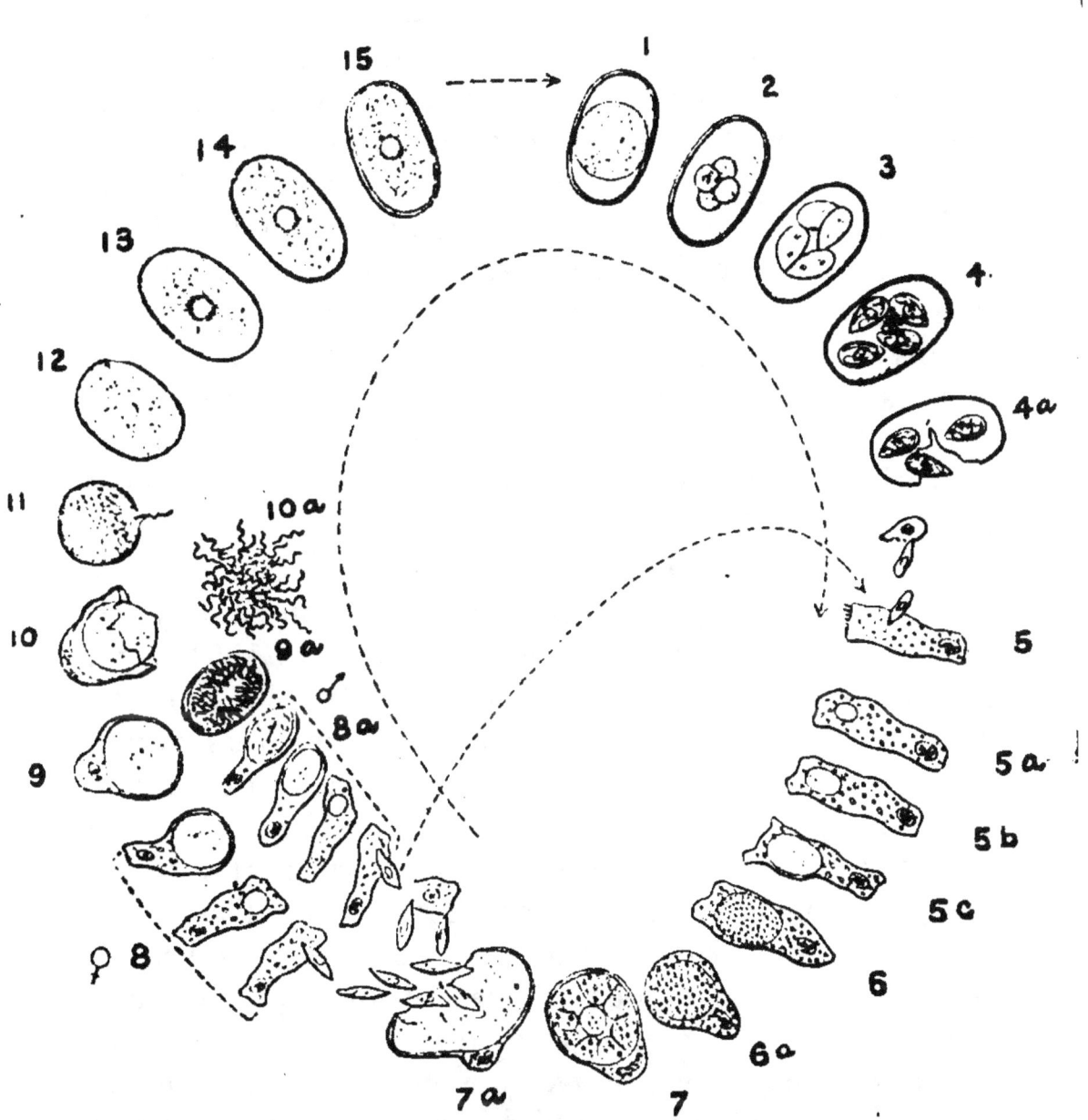

Fig. 8. — Diagrammatic representation of the life history of a coccidium. (After Cole and Hadley.)

then divides into two sickle-shaped sporozoids (cf. Fig. 8, 3 and 4). These sporozoids are then set free in the intestinal tract (4a) and each one penetrates with its pointed end an epithelial cell of the intestine as at 5. In the figure 5a, 5b, 5c, 6 and 6a, represent the succeeding stages of growth of the organism within the intestinal cell. As shown in 6a and 7, the parasite grows so large that it completely fills the cell. Finally these cells are broken down and torn off the intestinal wall. The stage of the parasite shown at 6a and 7 is known as the schizont. The next step is for the schizont to break up into a larger number of sharp pointed bodies as shown at 7a. These escape and enter other epithelial cells just as the somewhat similar bodies did at 5. At this point the organism may do one of two things. The small sporozoids from 7a may develop exactly as the sporozoids did from 5 to 7. This part of the life cycle, as shown by the shorter arrow from 7a to 5, may be repeated any number of times.

If, however, the conditions are not very good, *i.e.*, the bird is about to die, the sporozoids undergo an entirely different development, as shown at numbers 7 to 15 (Fig. 8). Here the sporozoids enter the epithelial cells and some develop into very large (egg) cells (female element), as shown in 8a. Others, 9a and 10a, form a very large number of minute motile zooids or sperms (male element) which unite with one or more of the large egg cells as shown at 11. After this sexual union there is developed the oöcyst like No. 1, with which we started. At all stages of this disease many of these cysts are carried to the outside with the feces and upon being picked up serve to infect other birds. Death is caused by the parasite attacking so many of the intestinal cells that the bird is no longer able to digest its food.

Other species of coccidia have different life cycles. Some are simpler and some more complex than the example given above.

Cole and Hadley [1] have advanced the claim that Blackhead (*Enterohepatitis*) is caused by a species of coccidium. This view is not admitted by others. (For discussion see pp. 94-99). Coccidia are, however, responsible for several serious diseases of poultry.

The question of the identity of species of coccidia attacking fowls, wild birds, and other animals has received considerable attention. Fantham [2] has shown that the coccidium of the grouse which causes the dwindling of the broods in the early summer is equally injurious to young fowls and pigeons. Some authors have claimed that the coccidium of the rabbit (*Eimeria* (*Coccidium*) *cuniculi*) is identical with that in birds. Fantham gives the results of experiments which show that the organism from rabbits is entirely distinct from that found in birds (*Eimeria avium*) and that the former will not cause disease in birds. A similar conclusion is reached by Jewett [3] except that he believes that under certain conditions the coccidium from rabbits can also produce disease in young chicks.

Diagnosis. — There are no special external symptoms of this disease until in an advanced stage. Adult fowls have considerable powers of resistance to this parasite, and the disease with them is more frequently seen in diarrhea, a chronic form. The symptoms are dullness, weakness, sleepiness, diarrhea, and loss of weight, although the birds retain their appetites for a considerable time. In many cases the only symptoms are diarrhea, with loss of weight,

---

[1] Cole, L. J., and Hadley, P. B., "Blackhead in Turkeys." Rhode Island Expt. Stat. Bul. No. 141, pp. 138-272, 1910.

[2] Fantham, H. B., "Coccidiosis in British Game Birds and Poultry." *Jour. Economic Biology*, Vol. 6, pp. 75-96, 1911.

—— "Experimental Studies of Avian Coccidiosis." *Proc. Zoöl. Soc.*, London, Vol. 3, pp. 708-722, 1910.

[3] Jewett, "Coccidiosis of the Fowl and Calf." *Jour. Comp. Path. and Therap.*, Vol. 24, pp. 207-225, 1911.

and after a time apparent recovery, though the germs continue to multiply in the intestine and to be spread with the droppings for several months afterwards. Fowls affected in this manner may die suddenly without previously showing any serious symptoms. Post-mortem examination often shows the liver enlarged and disfigured with whitish or yellowish spots. The ceca are inflamed and often clogged with pus and fecal matter.

Pigeons are affected with a more acute type of this disease in which the symptoms appear only a short time before death. Generally, however, they are dull and sleepy for a day or two, and sometimes they have a chronic form, characterized by diarrhea and loss of weight.

Geese with coccidiosis of the kidneys lose flesh rapidly, without apparent cause, and become very weak and almost unable to walk. They remain quiet most of the time, with belly resting upon the ground. Some of them are conspicuous by lying on their backs with their feet widely separated, and if placed upon their feet they take a few steps, fall, and resume their former position. In all such cases the birds lose their appetites and continue to get weaker until they die.

Treatment. — Medical treatment of coccidiosis is of very little avail in the present state of our knowledge. Isolation, cleanliness and disinfection are probably the most dependable treatment. Meyer and Crocker[1] claim that in an outbreak of coccidiosis in which nearly 1800 chickens died in from 3 to 6 weeks they had no success with any of the medical treatments used. They state that the epidemic was finally eradicated by isolation, disinfection and cleanliness.

[1] Meyer, K. F., and Crocker, W. J., "Some Experiments on Medical Treatment of Coccidiosis in Chickens." *Amer. Vet. Review*, Vol. 43, pp. 497–507, 1913.

Cole and Hadley [1] recommend for treatment of this disease in connection with blackhead in turkeys the following: "(1) Isolate the sick bird from the flock and place it in a dry, well lighted location free from cold and drafts. (2) Feed sparingly on soft, light, easily assimilable food, with little grain, especially corn." The chief preventive measures are to keep the birds on fresh ground; to isolate any birds showing the least sign of disease, to destroy all dead birds and to protect the birds from contamination carried either by new stock or by other poultry or by wild birds as sparrows, crows, etc.

According to Salmon [2] the most successful treatment has been to put 3 grains of copperas (sulphate of iron) to a quart, or 15 grains of catechu to a gallon of the water given the birds to drink. They should also be given an occasional dose of calomel ($\frac{3}{4}$ to 1 grain) or of castor oil (2 to 3 teaspoonfuls). They may also be given castor oil containing 5 to 10 drops of oil of turpentine with each dose.

Peritonitis

The thin serous membrane which lines the abdominal cavity and covers the internal organs is called the peritoneum. Inflammation of this membrane may occur in connection with the inflammation of certain internal organs such as the intestines, liver, kidneys, etc. In these cases the inflammation extends from the diseased organs on to the wall of the body cavity. Peritonitis may also be caused by the entrance of foreign bodies into the abdominal cavity. It may further be caused by severe bruises or injuries of the abdominal wall.

[1] *Loc. cit.*
[2] Salmon, D. E., "Important Poultry Disease." U. S. Dept. of Agric. Farmers' Bull. No. 530, pp. 1–36, 1913.

Peritonitis probably always follows the entrance of fecal matter into the body cavity through perforation of the intestines. Perforations may be caused by severe inflammation of the intestinal walls, or by the puncturing of the wall by parasitic round worms or other parasites, or by sharp pointed foreign bodies pushing through. Birds have a pernicious habit of picking up bright pieces of metal, glass, etc. Cases of peritonitis have occurred in the Maine Agricultural Experiment Station flock which were caused by the entrance of partly digested food from the gizzard through a perforation caused by a small nail, a watch spring or a pin.

By far the largest number of cases of peritonitis which have occurred in the Maine Station flock, however, have been associated with the failure of yolks to enter the oviduct or with the backing of partly formed eggs into the body cavity. Somewhat extensive studies [1] have shown that even when it is impossible for yolks to enter the oviduct the reproductive organs pass through their normal active cycles. The yolks are ovulated into the body cavity. Further if yolks can enter the oviduct, but if their passage is prevented at some level of the duct, either the partly formed egg remains in the duct forming immense masses (concrements) or they are carried back into the body cavity by antiperistalsis. These studies have shown that in about three-fourths of the experimental cases the birds are able to absorb these yolks

---

[1] Pearl, R., and Curtis, M. R., "Studies on the Physiology of Reproduction in the Domestic Fowl," VIII. On some Physiological Effects of Ligation, Section or Removal of the Oviduct. *Jour. Expt. Zoöl.*, Vol. 17, pp. 395–424, 1914.

—— "Studies on the Physiology of Reproduction in the Domestic Fowl," X. Further Data on Somatic and Genetic Sterility. *Jour. Exp. Zoöl.*, 1915.

Curtis, M. R., "Studies on the Physiology of Reproduction in the Domestic Fowl." XII. On an Abnormality of the Oviduct and its Effect upon Reproduction. Biol. Bul., Vol. XXVIII, pp. 154–163, 1915.

or eggs without any serious disturbance in their metabolic processes. Several cases have also occurred where supposedly normal birds were absorbing large numbers of yolks or eggs. These birds were in apparently perfect physical condition. The lumen of the duct was interrupted by fusion of the funnel lips; development of a tumor within the duct, rupture of the duct, or failure of a portion of the duct to develop.

Nevertheless in about one-fourth of the experimental cases, and in many natural cases of obstruction to the duct, death results from peritonitis, which is apparently caused by the failure of the peritoneum to resorb the yolks or eggs.

Diagnosis. — The sick birds appear restless and lose their appetite. There is a high fever. The abdomen is swollen, hot and tender. Pressure on the abdomen produces evidence of sharp pain. Usually, but not always, a severe thirst accompanies peritonitis. As the disease progresses the bird becomes weaker, is unable to stand and the legs are drawn up close to the body often with convulsive movements.

Post-mortem examination shows the peritoneum congested and covered with an opaque whitish or yellowish exudate. This gives it the appearance of being thicker than usual. In some cases quite large quantities of yellowish cheesy matter (pus) are formed. This may be in free lumps or masses or may adhere in a thin layer to the surface of the peritoneum. The abdomen sometimes contains a yellowish turbid serous liquid which may have an offensive odor.

Treatment and Prognosis. — Only very seldom is treatment for peritonitis successful. The disease is not usually recognized until in an advanced stage. Zürn[1] recommends wrapping parts of the bird in wet cloths and to give internally tincture of aconite, 2 drops (at the most) with a tea-

---

[1] Zürn, F.A., "Die Krankheiten des Hausgeflügel." Weimar, pp. 237, 1882.

spoonful of water 2 or 3 times a day. Sanborn recommends 1 grain opium pills twice a day to relieve pain, and warm liquid foods such as meat juice and milk in equal parts.

Abdominal Dropsy or Ascites

Etiology. — This disease is sometimes called chronic peritonitis. It is characterized by the accumulation of a large quantity of liquid in the abdominal cavity. In some cases the abdomen becomes so distended that it nearly or quite touches the ground when the bird is standing. Salmon says: "If examined by slight pressure of the hand the swelling is found to be soft and fluctuating; it will yield in one place and cause greater distention at another. That is, it gives the sensation of sac filled with liquid."

Abdominal dropsy may begin with a mild case of peritonitis which has continued for a long time without becoming serious. In young chicks it is said to be due to an anæmic condition produced by bad feeding and insanitary conditions. In older birds it may also result from this same cause or may be due to some obstruction of the venous circulation either by a tumor or by some structural disease of the abdominal organs.

Diagnosis. — The most marked symptom, of course, is the enlarged, flabby abdomen. Salmon says: "Fowls affected in this way are dull, disinclined to move, generally feeble with pale comb and diminished appetite."

Treatment. — "Treatment of this condition is not profitable, but in special cases, stimulating diet with considerable animal food, tonics and diuretics may be tried. Iodide of potassium or iodide of iron in doses of 1 grain is particularly indicated." (Salmon.) Tapping with a hollow needle or trocar through the skin and muscles of the abdomen and allowing the fluid to escape is also recommended. It will usually be found more profitable to kill the bird.

CHAPTER VI

Poisons

POULTRY on free range about farms and especially on small city lots often obtain poisonous substances. Most of the poisons obtained by fowls are the so-called mineral poisons. The chief symptom of poisoning by these substances is acute inflammation of the digestive tract. The narcotic or vegetable poisons on the other hand cause severe congestion of the blood vessels in the spinal cord and brain.

Among the principal poisons likely to affect poultry may be mentioned the following:

Common Salt, Nitrate of Soda, Concentrated Lyes

Common salt is most frequently obtained in excessive amounts from eating salt meat or fish. Suffram [1] reports a case in which fowls were poisoned by being fed a mash made of potatoes to which salt had been added. Milk and other liquids, prescribed after 13 had succumbed, resulted in the recovery of the 2 remaining. Chemical analysis of the food in the crops showed that each bird had taken from 10 to 14 grams of salt. From experiments instituted to determine the minimum toxic dose of salt it is concluded that a dose of 4 grams per kilogram (about $\frac{1}{15}$ oz. per pound) of body

---

[1] Suffram, F., *Rev. Gen. Médecin Veterinaire*, I. 13, pp. 698–705, 1909.

weight is sufficient to produce death. The fact that in these experiments one fowl resisted such a dose is thought to have been due to a certain toleration established by previous repeated injections of smaller doses. Zürn [1] gives a somewhat larger amount as fatal. He says that 15 to 30 grams (½ to 1 oz.) of common salt will kill a healthy hen in from 8 to 12 hours.

The writers had, some years ago, a rather serious experience with salt poisoning. In this case the salt was mixed with wheat, probably as a result of the latter following the former as a cargo in the hold of a vessel. A number of birds died, and the whole flock was made rather seriously ill before the cause was discovered.

Nitrate of soda is used as a fertilizer and is eaten by hens along with worms, etc., which they scratch up. Lye is obtained only when carelessly left about the grounds. The treatment for such poisons according to Salmon is to give "abundant mucilaginous drinks such as infusion of flaxseed, together with stimulants, strong coffee and brandy being particularly useful."

Arsenic may be obtained either from rat poison or from various arsenical sprays used to kill insects. *Copper* is used in such spraying mixtures as Bordeaux. Where spraying has been done properly there should be no danger of the birds getting enough of the poison to injure them. Sometimes, however, the vessels containing the mixtures are emptied within range of the fowls or the substances are handled carelessly in other ways.

The symptoms of arsenic poisoning are given by Beeck [2] as follows: "Secretion of large quantities of saliva, choking, hiccoughing, great anxiety and nervousness, little or no

---

[1] Zürn, F. A., "Die Krankheiten des Hausgeflügel."
[2] "Die Federviehzucht," 1908, p. 828.

appetite, thin, often bloody feces, slow and difficult breathing, unsteady walk, trembling and convulsions, expansion of the pupils. Death ordinarily occurs in a very short time." Treatment should be with sulphate of iron, calcined magnesia, or large quantities of milk. Salmon also recommends white of egg and flaxseed mucilage.

The special symptom of copper poisoning is diarrhea, the copper giving a blue or green color to the feces. Evidence of violent pain may follow with collapse, convulsions or paralysis. The circulation and respiration are weak. Usually fatal in a few hours. Large quantities of milk, white of egg, mucilage, and sugar water are recommended.

Lead and *zinc* poisoning occur chiefly from eating paint skins. The symptoms so far as they have been observed in poultry do not differ greatly from those seen in copper poisoning. The treatment recommended by Salmon is the same as for copper. With lead poisoning the sulphates of soda, potash or magnesia are recommended with the object of forming insoluble sulphate of lead.

Mercury Poisoning. — Mercury poisoning occurs chiefly through drinking bichloride (perchloride or corrosive sublimate) solution or eating mercurial ointment. The bichloride solution is a common antiseptic and is sometimes carelessly left where the birds have access to it. Ammoniated mercurial ointment is used to free the birds from lice (cf. p. 205). It is sometimes left where birds can get at the supply. More frequently poisoning results from the too free application of the ointment. If it is left in lumps on the feathers the birds will eat it. In man mercury poisoning is known to occur from too frequent or too long continued use of bichloride as a disinfectant, especially for large wounds. In the fowl it is not probable that such extensive treatment ever occurs. It is possible, though very unlikely, that poisoning due to ammoniated mercurial ointment may sometimes occur through the absorption from the skin. An excessive amount would have to be applied to cause such a result. As the mercury in this ointment is in an insoluble form it is much less likely to such absorption than is the bichloride.

The symptoms of mercurial poisoning are loss of appetite and frequent and sometimes bloody dysentery discharges from the bowels. At autopsy the mucosa of the whole intestinal tract is seen to be inflamed. The renal tubules of the kidneys are filled with a whitish crystalline deposit and the kidneys thus appear somewhat hypertrophied. Give large quantities of white of egg, milk, mucilage or flour and water.

Phosphorus may be obtained from rat poisons or from heads of matches. If large quantities of phosphorus are eaten by the bird severe inflammation of the stomach and intestine occurs and death results in from 1 to 2 hours. If only a small quantity is eaten the symptoms, according to Beeck, are weakness, languor, ruffled feathers, lack of appetite.

Strychnine is usually obtained by poultry from rat poisons. The distinctive symptoms here, according to Beeck, are the twisting of the spinal column and paralysis. The neck is twisted backward so that the head is often held over the rump. The treatment recommended by Beeck is to give inhalations of chloroform or internally 1 to 3 grains of chloral hydrate dissolved in 2 tablespoonfuls of water. The amount to be given depends on the size of the bird.

Ergot of Rye is one of the vegetable poisons which sometimes causes serious trouble among poultry. This is especially true in European countries. In this country so little rye is raised and fed to poultry that there is little chance for poisoning. The cause of the poisoning is a fungus which attacks the rye plants. The symptoms of ergot poisoning are trembling, intoxication, great weakness and gangrene of the comb, beak and tongue. The treatment is to give strong stimulants such as "brandy, coffee, camphor or quinine."

Fowls are occasionally injured by eating the leaves of poisonous plants. The sense of taste, however, protects the birds in most cases. Mr. H. B. Green [1] says in this connection: "Woodlands and fields abound in poisonous plants, and yet it is seldom, except in the case of birds that have been starved of green food and have become ravenous for it, that fowls ever succumb to vegetable poisons as thus obtained. Protection apparently lies in the fact that undesirable plants have repulsive flavors. Especially in suburban poultry keeping, danger arises when flower borders are weeded, seedlings thinned out, and plant rubbish swept up, if the resulting collection is thoughtlessly given to fowls in confined runs. Such birds are generally always ready for green food in any form and in their eagerness to satiate the craving the bad is often taken in with the good."

Ptomaine Poisoning

Fowls are subject to ptomaine poisoning. The cause of this is, of course, feeding spoiled or decayed food. Cases of this trouble are more frequent in small flocks where table waste is fed to a comparatively few birds.

Diagnosis. — The more common symptoms of ptomaine poisoning in fowls are: at first an unsteady gait showing lack of control (partial paralysis) of the muscles. If the birds are badly poisoned, prostration comes quickly. The birds lie in a relaxed condition with head and neck curled towards the breast. The comb turns black. In some cases there is a diarrheal discharge, occasionally bloody. Death usually occurs in a short time. In some respects the symp-

[1] *Illus. Poultry Record*, Vol. I, p. 689.

toms are similar to those of "limber neck" (see page 202).

Post-mortem examination shows a congestion of the liver, intestines and kidneys.

Treatment. — If the trouble is recognized in time the birds should be given a teaspoonful of castor oil. Follow this with sulphate of strychnine in doses of one-fifth grain every five hours.

Treatment for Poisons in General

In the great majority of cases a poisoned bird is not discovered until too late for treatment. Even if found it is usually not worth the poultryman's time to treat individual birds. The symptoms of the different poisons have been given in some detail with the hope that they may enable the poultryman to distinguish the kind of poisoning which they may encounter and may thus be able to remove the source of the trouble before other birds are affected. In the case of valuable birds the remedies indicated for the different poisons may aid in saving some of them.

CHAPTER VII

Diseases of the Liver

A LARGE number of diseases of the liver are described by writers on this subject. In the great majority of these diseases there are no external symptoms by which one can be told from another. The most common diseases which affect the liver may, for the moment, be divided into two rough classes which it is highly important for the poultryman to distinguish. These again can only be distinguished in dead birds, but the occurrence of cases of either kind in any number gives the poultryman a clew as to what the trouble may be and a chance to correct it. In the first of these two classes a post-mortem examination shows the liver covered with nodules of a cheesy-like appearance when opened. These nodules occur not only in the liver, but also in the spleen, intestine and other organs and sometimes in these latter regions without affecting the liver at all. With such symptoms we may be fairly certain that the trouble is tuberculosis and for a further discussion of this the reader is referred to Chapter IX.

In the second class of these diseases the liver is usually greatly enlarged, although in some cases it is shrunken and smaller than normal. With some of these diseases the liver may be spotted or marbled, but the condition is quite different from the cheesy nodules found in tuberculosis.

It is to this second class of diseases that the name "liver disease" properly belongs. "Liver disease" as popularly interpreted includes a number of different diseases distinguished by the pathologist. The more common are: Congestion of the Liver, Inflammation of the Liver, Atrophy of the Liver, Hypertrophy or Enlargement of the Liver, Fatty Degeneration of the Liver, and Jaundice.

The diagnosis of these different diseases is based entirely on the post-mortem appearances. In no one of them are there any outward symptoms which distinguish it from the others. Vale says it is impossible for the most scientific observer to diagnose either inflammation or congestion of the liver with positive certainty.

Further not only the symptoms, but also the causes and the treatments of these several diseases are essentially the same. The names of the diseases themselves indicate in a general way the post-mortem appearances.

For these reasons it seems best to give a brief discussion of the general causes of "liver disease" and the usual treatment. This will be followed by a brief account of each disease and its special symptoms and treatment, if any.

Cause of Liver Disease. — Lack of exercise and over-feeding, especially with rich albuminous foods, are the most common causes of diseases of the liver. In addition to these may be mentioned the obstruction of the circulation of the blood by disease of the heart and lungs. Congestion of the liver may be caused by any disease of the crop, gizzard or bowels that obstructs the circulation of the blood.

Undoubtedly the larger proportion of liver troubles results from improper feeding and housing. It is a common experience that complaints are more frequent in the latter part of the winter. The birds have been housed for some time without sufficient exercise and fed rich nitrogenous food. These causes operate slowly and since there are no outward symptoms of liver disease the poultryman is usually unaware of any trouble until his birds begin dying in the early spring. The conditions have then continued so long that it is often difficult to counteract them. This point emphasizes the necessity of keeping the flock under sanitary and healthful conditions.

Diagnosis of Liver Disease. — There are no special external symptoms. Some of the symptoms which often accompany these disturbances are: rough plumage, watery diarrhea, first brownish, then yellow; lack of appetite and indisposition to move. The comb may be purplish at first, becoming dark and then quite black. These, however, are all merely symptoms of disease in general that might apply to any one of a dozen or more ailments. The only certain method of recognizing the disease is by post-mortem examination. Every poultryman should be familiar enough with the normal appearance of the more important internal organs of a fowl to recognize abnormal appearances (see pp. 43–46). In general, when post-mortem examination shows the liver larger or smaller than normal, or congested with blood, or marbled, or spotted, we may assume that the bird probably had some form of liver disease. Of course, a diseased condition of the liver is often associated with other diseases, especially of the alimentary canal. Other organs should be examined in all cases to see if they are normal. Special care should be taken to distinguish tuberculosis from other diseased conditions of the liver and intestines.

Treatment. — Since it is not possible to recognize diseases of the liver by external symptoms, the treatment of individual birds is out of the question. If, however, post-mortem examinations show that a number of the birds are dying with liver trouble it is necessary to take some remedial measures regarding the entire flock.

The first thing that should be done is to change the diet. Less meat scrap and other nitrogenous food should be fed.

Less corn should be given and more green food added to the ration. The birds should be compelled to exercise more. If it is at all possible they should be gotten out of doors part of each day. They should have plenty of fresh air day and night.

These general remedial measures are only those which should be practiced at all times as a matter of general hygiene. When a flock has once become badly affected with some form of liver disease it cannot be expected that the changed conditions will remedy all of the trouble at once. The causes which have led to the diseased condition have been acting for a long time and it is only reasonable to expect that it will take some time to get the birds back into normal health again. Some birds will continue to become sick and die, even several weeks after the corrective measures have been put into operation. Robinson[1] advises disposing of the entire flock when they have been through a serious attack of liver disease and replacing them with healthy stock. This seems to be a more drastic measure than necessary unless the attack has been very bad indeed. Many other things, such as the value of the particular strain, the possibility of replacing the flock with as well bred birds, etc., should be considered.

With regard to the special diseases already mentioned the one most commonly met with, on intensive plants at least, is

Hypertrophy or Enlargement of the Liver

The cause of this trouble is chiefly concerned with food. In our climate it occurs most frequently towards the end of the winter. The birds have been confined to their houses most or all of the winter months. Very often they are over-

[1] Robinson, J. H., "The Common Sense Poultry Doctor." Boston, 1910.

crowded. The rich winter ration is continued after the weather begins to get warm and less heating food is needed. This combined with too little exercise and not enough green food favor indigestion and the accompanying sluggish action of gizzard and intestines. These are the immediate causes of trouble with the liver. It is said that feeding too much corn and barley is also responsible for much liver trouble.

Symptoms. — Mr. H. B. Green,[1] gives the following symptoms of hypertrophy of the liver. He believes this to be only a stage in the fatty degeneration of this organ. "The first sign that a fowl is tending towards fatty disease of the liver is increase in weight. The comb, wattles and face remain a bright red or take on a dull bluish tinge from congestion. This sign of sluggish circulation tells of full blood vessels, and explains how it is that apoplexy so frequently supervenes at this period. The excrement is an important symptom to note. It is generally at first semi-liquid, of a dark yellow color, and evacuations are frequent. Thirst is noticeable and a large quantity of water is drunk, especially after feeding. The appetite remains good, although the bird is capricious in what it eats. A post-mortem examination of a fowl in this phase of the disease will show a liver considerably enlarged, of a deep red color, engorged with blood, shining and greasy as though it had been soaked in oil, but fairly firm under the knife. The intestines are laden with masses of fat, so also are the mesentery, — or as it is termed by butchers, 'the leaf,' — the ovary and oviduct."

In the next stage "Diarrhea increases, the excrement perhaps bloodstained or blackened by congealed clots; the face, comb and wattles become a darker hue or if jaundice

[1] *Illustrated Poultry Record*, 1909, p. 691.

supervenes they may be pale or tinged with yellow bile; more fat is laid on internally and the liver will prove to be greatly enlarged. So large may this become by the deposit of fat globules between and in the substance of its cells that on one occasion I have removed from an Orpington cock a liver that turned the scale at a pound and a half. This stage is seldom passed and death usually takes place from syncope, or an accidental rupture of the softened liver."

Treatment. — Green says further: "Part of the treatment consists of a plentiful allowance of green food. Nothing in this way is better than freshly gathered dandelion leaves when procurable, for the taraxacum they contain is a valuable liver stimulant. It is not generally known that the sliced roots of the plants can be steeped in boiling water to make an infusion equally effective when the leaves are no longer obtainable. The roots should be gathered and stored in dry boxes. The infusion is conveniently mixed with the morning soft food and is always beneficial to birds in confinement as an occasional liver tonic."

Fatty Degeneration

As noted in the above paragraphs, Green regards this disease as a later stage in the hypertrophy of the liver. Salmon, on the other hand, believes it to be a quite different disease. The latter author says: "On *post-mortem* examination the liver is found shrunken, hardened and marbled or spotted with areas of grayish or yellowish tissue. A microscopic examination shows the liver cells to contain droplets of fat and the liver tissue degenerated and largely replaced by yellow fat globules.

As the disease is not recognized during life, treatment is out of the question. If a number of cases occur in the same flock, give greater variety of food and a run on the grass. In addition, bicarbonate of soda may be given in the drinking water to the amount of 1 or 2 grains a day for each bird."

Atrophy or Wasting of the Liver

This is very similar in many respects to the disease described by Salmon as fatty degeneration and probably arises from the same cause, *i.e.*, lack of variety in the food, especially lack of green food.

The post-mortem appearance and the treatment are the same as those given for fatty degeneration above. With both of the diseases a weekly dose of some laxative such as Epsom salts dissolved in water and mixed with the mash (a level teaspoonful to each bird) is to be recommended.

Congestion and Inflammation of the Liver

These are probably different stages of the same disease. The poultryman will find difficulty in distinguishing between this disease and that known as hypertrophy of the liver (cf. p. 90). The chief post-mortem difference is that in the latter disease the liver is more solid, not so easily torn or ruptured.

Diagnosis. — There are no external symptoms other than those of dullness and the general symptoms of disease. Salmon says: "It is difficult to make a diagnosis during the life of the bird. Post-mortem examination reveals a greatly enlarged liver engorged with blood, tender and easily torn or crushed."

Treatment. — Treatment of these diseases in individual birds is very rarely successful. The general treatment of the flock as recommended on page 89 should be attended to. The chief medicinal treatment should probably be frequent doses of Epsom salts.

Epsom salts together with bicarbonate of soda, 10 grains of each, given for 4 or 5 daily doses may be recommended also. This should be followed by the addition of a good tonic to the mash. (For stock tonic formula see p. 71.)

Jaundice

Jaundice or biliary repletion is said by Megnin to be due to long continued but moderate congestion of the liver. This leads to increased activity of this organ and is followed by the accumulation of a large quantity of bile in the gall bladder and ducts of the bird. This bile is absorbed by the blood vessels and causes poisoning which may lead to the death of the bird.

Diagnosis. — There are no specific external symptoms other than that the wattles and comb may be yellowish. This also occurs in other liver diseases. Post-mortem examination shows the gall bladder greatly distended with bile.

Treatment. — Give greater variety of food, especially more green food. Give Epsom salts frequently. Megnin recommends $\frac{1}{2}$ to 1 grain of aloes.

This completes the list of the liver diseases most commonly treated as such by poultry veterinarians. There are a number of other diseases which especially affect the liver or are caused by deranged function of this organ. These may most conveniently be mentioned at this place.

Blackhead (Infectious Enterohepatitis)

Blackhead is a contagious disease affecting the liver and intestines, especially the blind pouches or ceca of the latter. The disease is very quickly fatal among turkeys. The turkey is apparently more susceptible than any other bird to this disease. In certain portions of this country where once turkey raising was a promising industry it has practically vanished because of this disease. The disease is not usually as fatal to adult chickens but may cause very serious losses at times. It is believed by several prominent investigators of this disease that white diarrhea, so destructive to young chicks, is caused by the same organism as blackhead. (For further discussion of this see Chapter XVIII.)

The cause of blackhead disease according to Theobald Smith[1] is a minute parasitic protozoön known as *Amœba meleagridis*. These appear as minute round bodies not more than 10 microns ($\frac{1}{2500}$ inch) in diameter embedded in the submucous and intramuscular tissue of the wall of the ceca and may extend even beyond these to the mesenteries. In the liver there are circular spots (Fig. 9) representing partial necrosis of the liver tissue and in these spots the same organisms are also present in great numbers. The analogy between this organism and that concerned in human amœbiasis is very close.

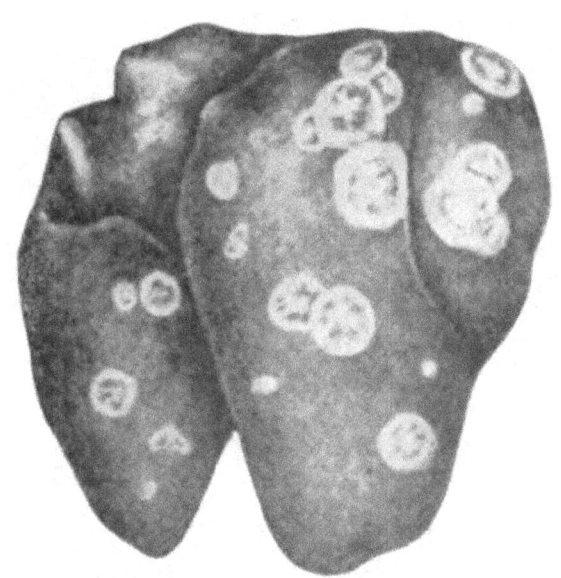

Fig. 9. — Showing condition of liver in "blackhead." (Modified after Moore.)

More recently Cole and Hadley[2] at the Rhode Island

---

[1] Smith, Theobald., "An Infectious Disease among Turkeys Caused by Protozoa (Infectious enterohepatitis)." U. S. Dept. of Agr., Bur. Anim. Ind., Bul. 8, pp. 7–38, 1895.

[2] Cole, L. J., and Hadley, P. B., "Blackhead in Turkeys." Rhode Island Expt. Stat. Bul. No. 141, pp. 138–272, 1910.

Experiment Station have claimed that the causative organism belongs to another group of protozoa known as *Coccidia*. The point to the discussion as to the cause of this disease lies in the fact that the *Coccidium* has a very different life history (cf. p. 73) from the *Amœba*, consequently it has different methods of dispersal and different means must be used in combating it.

Cole and Hadley claim that the amœba described by Smith is one of the stages in the life history of the coccidium. Smith, however, in a more recent paper [1] reaffirms his position with regard to the amœbic cause of the disease. He claims that the Rhode Island authors have confused the fact that coccidia are frequently present in birds as an entirely separate infection. He states that there is "ample evidence to show that enterohepatitis may run its course in a flock without the presence of a single coccidium cyst to suggest coccidiosis. It is evident that coccidiosis among birds has been frequently seen during the past 30 years but without involvement of the liver."

These criticisms by Smith were partially answered by Cole and Hadley [2] but the chain of evidence presented by them is far from complete. Not until the complete life history of the coccidium has been worked out will there be conclusive evidence as to whether or not it is concerned in this disease. In the meantime it appears that the contention of Smith is well founded, viz., that the amœba and the coccidium are separate entities and that the latter when present is only a secondary infection.

The method of infection by the amœba is as follows, the account being based upon that given by Salmon (*loc. cit.*):

[1] Smith, Theobald, "Amœba meleagridis." *Science*, N. S., Vol. 32, pp. 509–512, 1910.

[2] Cole, L. J., and Hadley, P. B., "Amœba meleagridis." *Science*, N. S., Vol. 32, pp. 918–919, 1910.

The amœba leaves the bodies of the sick birds with the excrement and infects other birds by entering the digestive organs with the food and drink. It passes along the alimentary canal until it arrives at the two blind pouches or lateral extensions called the ceca (Fig. 7), where it begins its growth and produces the first signs of disease. Here it penetrates the lining membrane, increases rapidly in numbers, and sets up an inflammatory process which leads to a great thickening of the intestinal wall and to the filling up and obstruction of the tube with an accumulation of yellowish white or grayish cheesy material that is deposited in concentric layers.

The changes which are almost constantly found in the liver are explained by assuming that the microbes are carried by the blood from the diseased ceca to the liver, and are there deposited at different points, where they multiply and spread in all directions. In this way are formed the numerous centers of disease which appear on the surface of the liver as yellowish spots, but which when cut across are seen to be irregularly spherical in shape. The amœbæ are liberated in large numbers both in the ceca and in the liver, are mixed with the intestinal contents, and are distributed with the droppings.

There is some difference of opinion as to whether the amœba is ever present within the egg of diseased turkeys, but the indications are that the infection is not carried in this way. It no doubt often exists on the outside of the shell, from contamination when the egg passes through the cloaca, and for this reason the eggs should be carefully cleaned before they are put under the sitting hen or into an incubator.

An important recent conclusion is that common fowls harbor this parasite, although they rarely suffer sufficiently from its attacks to show marked symptoms of disease. They scatter the contagion constantly, however, and young turkeys, being more susceptible, contract a fatal form of the disease and are nearly all destroyed by it. For this reason it is very difficult to raise turkeys on or near grounds where there are common fowls.

Diagnosis. — The symptoms of blackhead are most frequently seen in young turkeys, commonly called "poults," which are from 2 weeks to 3 or 4 months old. When young poults are infected experimentally by feeding them diseased livers they usually die in about two or three weeks, but when infected naturally they generally take in a smaller quantity of contagion and live a longer time.

The affected birds at first appear less lively than usual, are not so active in searching for food, and when fed show a diminished appetite. Diarrhea is a nearly constant symptom, being due to the inflammation of the ceca. As the disease progresses there is more dullness and weakness, the wings and tail droop, and there is often the peculiar discoloration of the head which led to the disease being called "blackhead." There is increasing prostration and loss of weight; the affected birds, instead of following their companions, stand about in a listless manner, indisposed to move and paying little attention to what occurs about them.

The greater part of the affected poults die within three or four months after hatching; but with some the disease takes a more chronic form and does not cause death for a year or more. Nearly all die sooner or later from the effects of the disease, but in a small proportion of the cases there is healing and recovery.

The finding after death, in young turkeys, of the diseased and thickened ceca, plugged with cheesy contents, together with the yellowish or yellowish-green spots in the more or less enlarged liver are sufficient indications to warrant a diagnosis of blackhead.

Treatment. — The treatment of diseased birds has not given satisfactory results. The remedies most often used are sulphur 5 grains, sulphate of iron 1 grain; or benzonaphthol 1 grain, salicylate of bismuth 1 grain; or sulphate of iron 1 grain, salicylate of soda 1 grain. These remedies should be preceded and followed by a dose of Epsom salts (10 to 35 grains), or of castor oil ($\frac{1}{2}$ to 3 teaspoonfuls). Fifteen grains of catechu to the gallon of drinking water may also have a beneficial effect. It seems clear, however, that it does not pay to doctor the sick poults and that the only hope of success at present is in preventing their infection.

The measures of prevention which have been suggested are (1) obtaining eggs from birds believed to be healthy; (2) wiping the eggs with a cloth wet with alcohol (80 to 90 per cent) before they are placed in the incubator or under the hen for hatching, to remove any contagion that might be on the shell; (3) hatching in an incubator, or at least removing the eggs from under the hen a day or two before hatching would occur, wiping with alcohol, and finishing in an incubator, in order to avoid exposing the poults to the hen; (4) placing the young poults on the ground at a distance from all other domesticated fowls and which has not recently been occupied by other fowls; (5) excluding so far as possible pigeons, other wild birds, and rats and mice from the houses and runs occupied by the turkeys; (6) the frequent disinfection of the houses, feed troughs, drinking fountains, etc.; (7) the immediate killing of diseased birds and the destruction of their bodies by fire.

These radical measures are necessary, and in sections of the country which are not too intensely infected they will make it possible to carry on the turkey industry successfully. However, it must be admitted that up to the present blackhead has proved to be one of the most difficult of all diseases to prevent or eradicate.

The destruction of the contagion, after it has been introduced into a poultry yard, has also been found difficult or impossible. Some have proposed to dip up and burn the surface soil to a depth of several inches, which might be done with small yards but is impossible with large ones. In most cases the poultryman must be contented with the application of a layer of freshly burned lime that has been carefully slaked to a fine, dry powder. After a few weeks this ground should be plowed and another layer of lime applied. The manure which has accumulated should be burned or mixed with lime and plowed into the ground of some distant fields. The wall and floors of the buildings should be covered with a good limewash containing 6 ounces of carbolic acid to the gallon. The fences should receive a coat of limewash. The feeding troughs and drinking vessels should be put into a kettle of boiling water for half an hour. Troughs too large for this should be burned and replaced by new ones. After these measures are adopted, the longer the premises are left vacant the more likely is the contagion to be completely destroyed. The freezing and thawing of a winter and spring will be found of great assistance. In beginning with a new flock the precautions already mentioned must be adopted to prevent the infection of the premises.

Cercomoniasis

This is frequently called "spotted liver." It, like many other liver diseases, is associated with intestinal trouble, especially severe diarrhea, that attacks poultry during the summer months. The disease is caused by a flagellate micro-organism known as *Monocercomonas gallinarum*. The post-mortem appearance of the liver in this disease shows usually slightly depressed yellowish necrotic areas or spots.

This fact usually distinguishes this disease from tuberculosis where there are prominent rounded cheesy nodules. In pigeons, however, this cercomonad is said to cause rounded prominent nodules about the size of a pea.

This same organism (*Monocercomonas gallinarum*) is also said to be responsible for other diseases. The most important of these is one form of roup. Canker in squabs and intestinal diarrhea in poultry are other diseases attributed to this parasite.

This disease can be held in check, it is said, by keeping the poultry plant well cleaned and disinfected and by giving the birds an occasional purgative, *e.g.*, Epsom salts.

In *aspergillosis*, the liver often presents the appearance of being "studded all over with minute, whitish or yellowish spots." This disease is discussed in Chapter XI.

Gout

In cases of visceral gout the liver and adjoining organs are covered with a fine chalky sediment. This substance consists of crystals of urate of soda. (See Chapter XIV for detailed description.)

Sarcomatosis and Carcinomatosis

In some cases the liver is affected with tumors or cancers. These are usually found in connection with similar developments on the ovaries (see Chapter XX).

CHAPTER VIII

Fowl Cholera, Fowl Typhoid and Fowl Plague

Cholera

Fowl cholera is a virulent, usually fatal and highly infectious disease. It is entirely distinct from the ordinary forms of enteritis with which it is often confused by poultrymen. Fowl typhoid and infectious leukæmia are also often mistaken for cholera. Genuine fowl cholera is rather rare in this country but is much more common in Europe. According to some investigators it is now on the increase in this country. This disease was first reported in this country about 1880 by Salmon (Rept. U. S. Comm. of Agric.). Owing to the lack of proper bacteriological methods at that time Salmon was not able with certainty to identify this disease with the European cholera. From certain experimental work he concluded that some of the symptoms exhibited by the disease in this country were different from those described by European writers. About 1894 Moore [1] obtained material from several outbreaks of supposed cholera but found this disease to differ in some important respects from the European trouble. Later Curtice [2] described a disease similar to that of Moore's under the name of fowl typhoid.

---

[1] Moore, V. A., "A Study of a Bacillus Obtained from Three Outbreaks of Fowl Cholera." U. S. Dept. of Agric. Bur. Anim. Ind., Bul. No. 8.

[2] Curtice, R., "Fowl Typhoid." Rhode Island Agr. Expt. Stat. Bul. No. 87.

Fowl Cholera, Fowl Typhoid and Fowl Plague   103

What appears to be the genuine European fowl cholera has been reported several times within the last few years.

Etiology. — Fowl cholera is caused by a minute bacterium

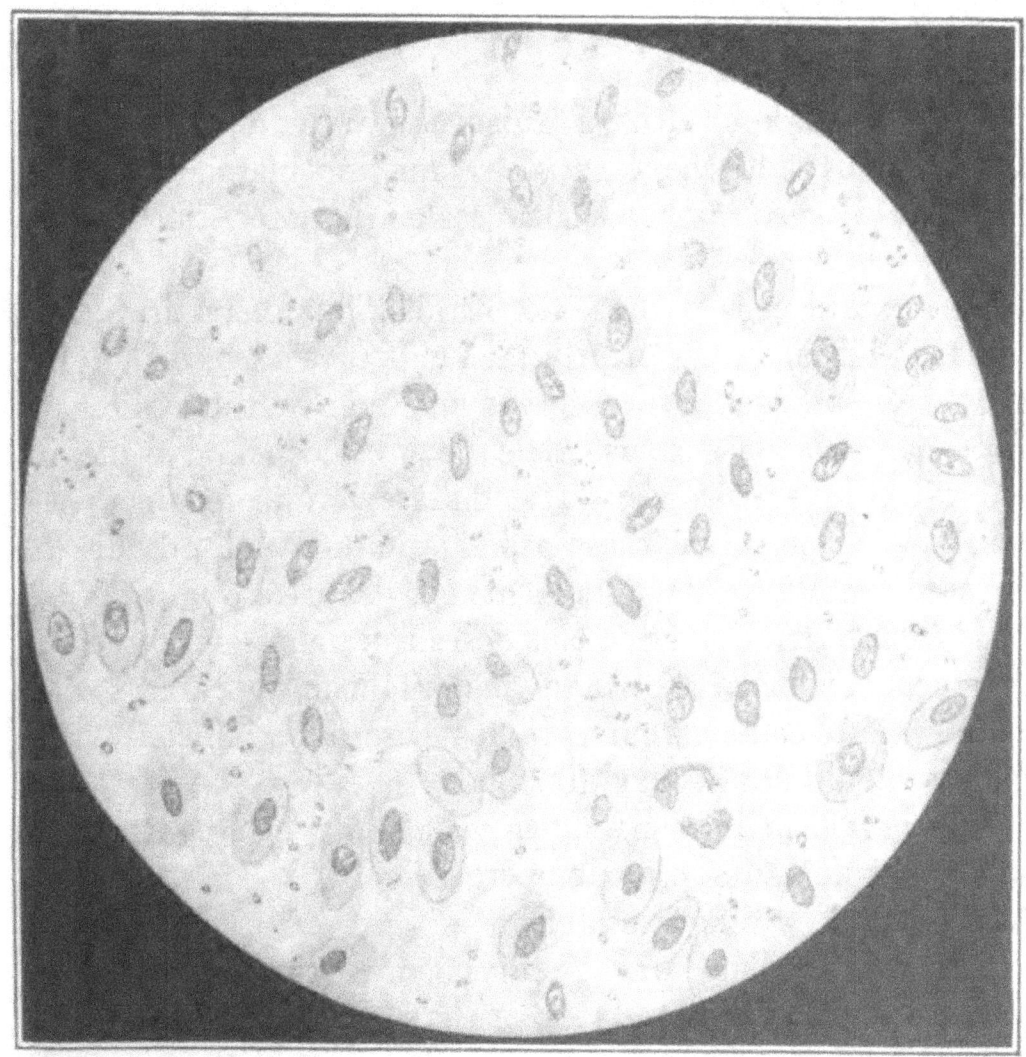

Fig. 10. — Stained preparation of the heart blood of a pigeon infected with fowl cholera. The small objects between the blood corpuscles, each showing two dots of color, are the cholera bacteria. (After Kolle and Hetsch.)

known as *Bacillus avisepticus* (*B. bipolaris septicus*). This is a small oval organism which when stained and placed under the microscope shows a dot of color at each end while the middle part remains entirely unstained (Fig. 10).

The organism was first recognized about 1878. In 1880 Pasteur [1] used it in his epoch making work on the production of immunity with attenuated or non-virulent cultures. This organism belongs in the same group as the hog-cholera bacillus (*B. suisepticus*), rabbit septicæmia and many other destructive bacteria.

In the diseased bird the organism can be found in the blood and in nearly all the organs. When inoculated into the breast of a pigeon or fowl it causes a characteristic hemorrhagic swelling.

This organism is pathogenic for all kinds of poultry and domestic birds and for nearly all kinds of wild birds. It is also very pathogenic for rabbits and many other animals. For larger animals, such as cattle, horses, sheep, and swine, it will cause severe or even fatal disease if injected intravenously. When fed to these animals it does not produce a diseased condition. Dogs and cats can eat great quantities of meat from birds dead from this desease without experiencing any inconvenience. For man this organism appeared to be pathogenic to some extent. In any case birds infected with cholera should not be used for food.

This organism is easily destroyed by drying, by the ordinary disinfectants, by a temperature of 132° F. for fifteen minutes, and by direct sunlight.

Diagnosis. — The earliest indication of the disease is a yellow coloration of the urates, or that part of the excrement which is excreted by the kidneys. This in health is a pure white, though it is frequently tinted with yellow as a result of other disorders than cholera. While therefore this yellowish coloration of the urates is not an absolutely certain proof of cholera, it is a valuable indication when the disease has appeared in a flock and an effort should be made to check

---

[1] Pasteur, L., "Sur le cholera des poules et l'attenuation du virus du cholera des poules." Comptes Rendus, 1880.

its course by isolating the sick birds as soon as possible. In regard to the yellow or green excreta Hadley[1] says: "This is a very characteristic symptom. The excrement of normal fowls is not yellow; and when it is green it is a dark green, approaching black. In cholera both yellow and green are bright; the green is often an emerald green. These different colors may occur either together or separately and both are usually accompanied by diarrhea and thick mucus. In case it is known that cholera is in the neighborhood, it is well for a poultryman to examine, from day to day, the character of the droppings on the dropping board."

In other cases the first symptom is diarrhea in which the excrement is passed in large quantities, and consists almost entirely of urates mixed with colorless mucus. Generally the diarrhea is a prominent symptom. The excrement is voided frequently, and consists largely of urates suspended in a thin, transparent, sometimes frothy mucus. The urates have a deep yellow color, which in the later stages of the disease may change to a greenish cast.

Soon after these first symptoms appear the bird separates itself from the flock, the feathers are roughened or stand on end, the wings droop, the head is drawn down towards the body and the general outline of the bird becomes spherical or ball shaped. At this period there is great weakness, the affected bird becomes drowsy and may sink into a deep sleep which lasts during the last day or two of its life and from which it is almost impossible to arouse it. The crop is nearly always distended with food and apparently paralyzed. There is in most cases intense thirst. If the birds are aroused and caused to walk there is at first an abundant discharge of excrement followed at short intervals by scanty evacuations.

The disease may be acute, in which case the bird dies in

[1] Hadley, P. B., "Fowl Cholera and Methods of Combating It." Rhode Island Agr. Expt. Stat. Bul. 144, pp. 309–337, 1910.

from a few hours to a day or two. Or it may be subacute, in which case the bird lingers for several days. Likewise within the flock the birds may die rapidly until the majority of the flock are gone within a few days or they may die a few at a time throughout a period of several weeks. The period of incubation, *i.e.*, from the time of exposure until the first symptoms, varies from 1 to 2 days in geese and from 4 to 9 days in chickens.[1] Salmon states that the incubation period may be as much as 20 days.

Examination of the dead birds shows inflammation of the digestive organs, kidneys, and mesenteries in nearly all cases. According to Ward[2] "punctiform hemorrhages are found upon the heart with almost absolute uniformity. The liver is very frequently marked with punctiform whitish areas." Sections show that the areas of necrotic tissue are present throughtout the liver tissue. The blood vessels of the liver are congested. According to Ward the next most striking lesions are found in the reddened and bleeding mucosa of the first and second folds of the small intestine (next to the gizzard). These reddened areas can even be seen from the outside of the intestine. The intestinal contents are either a cream colored pasty mass or may be brownish or even green in color. "Lesions are very rarely observed in other portions of the intestine. The ureters are noticeable in practically all cases by reason of the yellow-colored urates that they contain. The nasal cavity, pharynx and oral cavity frequently contain a viscous mucous fluid, probably regurgitated from the crop."

Mode of Transmission. — The manner in which this disease gains admission to an apparently healthy flock is often a puz-

---

[1] Ostertag, R., und Ackermann, P., "Zeitschr. Infektkr. u. Hyg. Haustr." Bd. 1, pp. 431–441, 1906.

[2] Ward, A. R., "Fowl Cholera." Cal. Agr. Expt. Sta. Bul. 156, pp. 3–20, 1904.

zling question. Thus while many epidemics owe their origin to the importation of infected birds, birds returning from poultry shows, or to the presence of infected wild birds, some epidemics appear to arise spontaneously. Recently it has been found that the causative organism is occasionally present in the intestines of an apparently healthy bird. These spontaneous epidemics are probably to be explained as due to the increase in virulence of such organisms. After passage through two or three hens this virulence is still further increased so that an epidemic is started. Later on this same strain may decrease in virulence but may remain in the flock only to break out again a year or two later.

Within the flock the infection is generally transferred through the food or drinking water contaminated with the excrement of sick birds. It is also possible for birds to be infected through wounds of the skin or by inhalation of the germs in the form of dust suspended in the air. In other cases the dissemination of the disease is undoubtedly due to the fowls eating the dead bodies of infected birds.

Treatment. — At the present time there is no certain cure known for fowl cholera after the bird has been infected. While some birds may recover of their own accord it is probable that such birds are a source of danger to the flock for some time afterwards. Müller[1] states that infected fowls continue giving off the bacteria in the urates three weeks after infection, and that the organs contained virulent material after a period of six months.

Under ordinary circumstances, if it is known that fowl cholera is on the premises, every bird showing marked symptoms of this disease should be killed at once. The birds should be killed in such a manner that their blood will not be spilled near the houses or runs. Every drop of blood from

[1] Müller, J., "Monatsh. Prakt. Tierheilk." Bd. 21, pp. 385–413, 1910.

an infected fowl contains thousands of these bacteria and may serve to infect other birds. The bodies of the birds should be burned, or if this is not possible they should be buried deeply so that dogs and other animals will not dig them up.

A pest house should be established as soon as it is known that the disease is present. This should be located at some distance from the regular houses. Every bird showing the slightest symptoms of the disease should be removed to this house at once. All the litter and droppings from the regular houses and runs should be scraped up and burned and everything about the place thoroughly disinfected. Spray the houses with a good disinfectant. Do not use any litter in the houses unless a light coat of sawdust and this should be replaced by fresh every day. The runs and yards should be thoroughly disinfected and should be plowed often.

If these measures are carried out with conscientious attention to details it is a relatively simple matter to eradicate this disease. However, the disinfection must be kept up for some months after the birds have ceased to die. Otherwise the infection may return.

In connection with his work on an outbreak of fowl cholera in California, Ward points out the following important conclusion: "Cholera and other infectious diseases may exist in a fowl in a sort of inactive chronic condition and there is no doubt concerning the agency of such a case in spreading the disease. Thus, fowls not suspected of being diseased may have the disease smoldering among them. The fact that occasionally a single fowl dies of cholera means that a severe loss may occur at any time."

The practical recommendation for an outbreak of fowl cholera then is to kill and destroy all sick birds, confine all well birds to small runs. Disinfect these runs and the houses daily. After the outbreak is over and the birds have ceased dying it is best to market all flocks in which the cholera appeared. This latter precaution will often prevent a second outbreak some months later.

Methods of prevention are always the most satisfactory. The careful poultryman will guard his flock against all infectious diseases by methods of quarantine, disinfection and general cleanliness. At the same time the birds should be fed to keep them in the best of health. On these points read Chapter II.

A large amount of work has been done upon remedies and preventives for this disease. Recently Hadley[1] has recommended the subcutaneous injections of 5 per cent carbolic acid as a treatment for individual birds. This author says: "At the Rhode Island Station attempts have been made to prevent the development in fowls of cholera artificially produced by inoculation with the fowl cholera organism. The protective inoculations have involved subcutaneous inoculations with a 5 per cent solution of carbolic acid in amounts of from 2 to 4 c.c. daily.

"The results thus far secured show that the inoculations as given protected artificially infected birds, and did no harm to birds that were in normal health. They therefore suggest that subcutaneous inoculations with carbolic acid have a protective and perhaps a therapeutic value in fowl cholera."

Much work has also been done, especially in Europe, upon methods of protective inoculation against this disease. The best success has been obtained by the use of immune sera. Such a serum is prepared by immunizing a large animal, horse or cow, by repeated injections of this organism. The serum from this animal is then collected and used to inoculate healthy birds. Such an immune serum gives a passive im-

[1] Hadley, P. B., "Fowl Cholera and Methods for Combating It." Rhode Island Agr. Expt. Stat. Bul. 144, pp. 309–337, 1910.

munity to the bird which will last about 18 days [1] after each injection. The fact that such immunity is not permanent renders it of little value in treating an infected flock. Lisoff [2] reports the use of such a serum in a large number of epizoötics (3876 birds) and states that the disease can easily be held in check. As a curative agent he says the figures show a reduction in mortality from 90 to 22 per cent where the serum was used.

Such protective serum is largely used in Denmark and other countries for treating geese and other birds which are being shipped into the country.

Other methods of producing immunity against this disease have also been tried. These involve the injection of dead cultures or of living avirulent cultures or of the sterilized exudate obtained by injecting cultures into the pleural cavities of other animals. In the main these methods have not proven very successful in a practical way. For instance the majority of avirulent cultures will not produce immunity against all virulent strains. This whole question is now being studied by the Rhode Island Experiment Station.[3]

---

[1] Kitt, T., "Monatsh. Prakt. Tierheilk." Bd. 16, pp. 1–19, 1904.

[2] Lisoff, P. W., ["Anti-fowl-cholera Serum and Its Practical Significance"]. (Russian) Vet. Nauk (St. Petersburg), Bd. 40, pp. 804–818, 1910.

[3] For example see:

Hadley, P. B., and Amison, E. E., "A Histological Study of Eleven Pathogenic Organisms from Cholera-like Diseases in Domestic Fowls." Rhode Island Agr. Expt. Stat. Bul. 146, pp. 43–102, 1911.

Hadley, P. B., "The Rôle of Homologous Cultures of Slight Virulence in the Production of Active Immunity in Rabbits." Rhode Island Expt. Stat. Bul. 150, pp. 81–161, 1912.

—— "The Reciprocal Relations of Virulent and Avirulent Cultures in Active Immunization." Rhode Island Expt. Stat. Bul. 159, pp. 383–403, 1914.

Fowl Typhoid

In 1895 Moore [1] described a disease of fowls caused by an organism which he named *Bacterium sanguinarium*. He called the disease infectious leukæmia owing to the fact that it is accompanied by a marked increase in the number of white blood corpuscles. This disease is discussed in detail in Chapter XII.

At the time of his original description of this disease Moore pointed out that it was frequently mistaken for fowl cholera, but he called attention to a number of specific differences (see p. 188). He also says that the organism causing this disease closely resembles in its physiological properties *Bacillus typhosis*, the cause of human typhoid. In more recent literature this disease has frequently been called fowl typhoid. Smith and Ten Broeck [2] have pointed out that this organism has many diagnostic features in common with the human typhoid bacillus. Even in its agglutination reactions it closely resembles the typhoid organism. The fowl organism differs from the human, however, in being non-motile.

In spite of the marked resemblance the two organisms are apparently distinct. Mitchell and Bloomer [3] state that the chicken is highly resistant to the human typhoid organism. In the experiments reported the chickens failed either to contract the disease or to act as a carrier. The experiments involved both feeding the organism and injecting it intrave-

---

[1] Moore V. A., "Infectious Leukæmia in Fowls — a Bacterial Disease Frequently Mistaken for Fowl Cholera." U. S. Dept. of Agr. Bur. of An. Ind. Repts., 1895 and 1896, pp. 185–205.

[2] Smith, T., and Ten Broeck, C., "Agglutination Affinities of a Pathogenic Bacillus from Fowls (fowl typhoid) *Bacterium sanguinarium* Moore) with the Typhoid Bacillus of Man." *Jour. of Medical Research*, Vol. 31, pp. 503–521, 1915.

[3] Mitchell, O. W. H., and Bloomer, G. T., "Experimental Study of the Chicken as a Possible Typhoid Carrier." *Jour. of Medical Research*, Vol. 31, pp. 247–250, 1914.

nously. These experiments are not extensive enough to prove absolutely that chickens cannot become typhoid carriers.

Pfeiler and Rehse [1] have shown that while the fowl typhoid organism (which they renamed *B. typhi gallinarum alcalifaciens*) is extremely virulent for chickens, it does not attack ducks, geese or pigeons.

In another recent paper Smith and Ten Broeck [2] have shown that the fowl cholera organism produces a toxin that is very poisonous to rabbits. They suggest that possibly this same organism may play a part in the food or so-called ptomaine poisoning in man.

In still another paper Smith and Ten Broeck [3] have shown that the fowl typhoid organism shows many points of resemblance to *Bacillus pullorum*, the cause of white diarrhea in young chicks (cf. p. 295). It is only by certain fermentation tests that the two can be distinguished.

A further discussion of this disease together with recommendations for prevention are given under infectious leukæmia on pages 186-189.

Fowl Plague

This disease is to be sharply separated from fowl cholera with which it is often confused. So far as the writers are aware this disease has never appeared in the United States. It is by no means uncommon in Europe. In spite of the fact

---

[1] Pfeiler, W., and Rehse, A., "Bacillus typhi gallinarum alcalifaciens." Mitt. Kaiser Wilhelms Inst. f. Landwirtschaft, Bromberg, Bd. 5, pp. 306-321, 1913.

[2] Smith, T., and Ten Broeck, C., "The Pathogenic Action of the Fowl Typhoid Bacillus with Special Reference to Certain Toxins." *Jour. of Medical Research*, Vol. 31, pp. 523-546, 1915.

[3] Smith, T., and Ten Broeck, C., "A Note on the Relation between *B. pullorum* (Rettger) and the Fowl Typhoid Bacillus (Moore)." *Jour. of Medical Research*, Vol. 31, pp. 547-555, 1915.

that considerable work has been done upon fowl plague, comparatively little is known about it. The following notes are gathered from such literature as is at hand.

Etiology. — No definite organism has ever been isolated in connection with this disease. Depperich [1] stated (1907) that all the then available evidence indicated that it is caused by an ultra-microscopic, filterable virus. Russ [2] states that the blood from cases of this disease is extremely virulent, being fatal when given in such extreme dilution as 1 to 1,000,000,000. The virus of this disease appears to be in some way attached to or included in the red blood corpuscles. By centrifuging out these corpuscles it is possible to remove a large portion of the virus from the blood. Landsterner [3] performed certain experiments which indicated that the causative organism may be a protozoön associated with the blood corpuscles. In this respect the causative factor in the disease appears to show some resemblance to the filterable virus of hog-cholera, according to recent work.[4]

Diagnosis. — Fowl plague is known to affect chickens, turkeys, guinea-fowl, geese, pheasants, and many wild birds. The lesions of the disease resemble those produced by phosphorous poisoning.[5] The surface of the heart may be covered with small blood clots (ecchymoses). It can be distinguished from fowl cholera by the presence of hemorrhages under the epicardium and an exudate in the pericardial cavity.

---

[1] Depperich, C. H., *Fortsch. Vet. Hyg.* Bd. 4, pp. 217–250, 1907.

[2] Russ, V. K., *Arch. Hyg.* Bd. 59, pp. 286–312, 1906.

[3] Landsterner, K., *Centralb. f. Bakt.*, etc., Abt. 1, Bd. 38, pp. 540–542, 1906.

[4] For example:
King, W. E., and Hoffman, G. L., "Studies on Hog Cholera — *Spirochæta suis*, Its Significance as a Pathogenic Organism." *Jour. Infec. Dis.*, Vol. 13, pp. 463–498, 1913.

[5] Freese, *Deut. Tierarztl. Wchnschr.* Bd. 16, pp. 173–177, 1908.

Marchoux[1] claims to have shown that the virus is not transmitted through the feces but that it is probably transmitted by some mite or tick. Experiments by others have failed to prove definitely that it is transmitted by such parasites.

Several investigators have called attention to an apparent relationship between this disease and rabies. Rosenthal states that subdural inoculation of fowls with the virus of this disease produces death with violent symptoms resembling rabies. Schiffmann[2] states that in the cerebrum of artificially inoculated geese certain corpuscles are found which in some respects resemble the Negri bodies of rabies. The two, however, are not identical.

Control. — The methods for the control of this disease must be similar to those of cholera. Sick birds must be isolated or killed and great care taken that the blood of infected birds is not spilled in the houses or yards.

[1] Marchoux, E., *Compt. Rendus Soc. Biol.* T. 68, pp. 346–347, 1910.

[2] Schiffmann, J., *Centbl. f. Bakt.*, etc. Abt. 1, Bd. 45, pp. 393–403, 1907.

CHAPTER IX

Tuberculosis

Tuberculosis in fowls has long been a serious pest in Euorpe. Zürn in his "Krankheiten des Hausgeflügels," published in 1882, devotes several pages to the description of this disease as it occurred in Germany. Its appearance in this country, however, seems to have been much more recent.

Salmon, whose book was published about 1888, says that the disease "is by no means rare in the United States if the statements of our professional men are to be accepted." However, at that time very little had been done in the way of bacteriological diagnosis and no doubt many of the early reports were unreliable.

The disease was first reported on the basis of bacteriological examination in 1900 by Pernot.[1] In 1903 Moore and Ward[2] reported investigations on avian tuberculosis in California. They found "a number of flocks in which the mortality from the disease was very high." Fowl tuberculosis was reported from western and central Canada in 1904 by Dr. C. H. Higgins.[3] In 1906 it was reported from New York and in 1907 from southern Michigan. The

---

[1] Pernot, "Investigation of Disease of Poultry." Oregon Agr. Expt. Stat. Bul. 64, 1900.

[2] Moore, V. A., and Ward, A., "Avian Tuberculosis." *Proc. Am. Vet. Med. Assoc.*, St. Paul, 1903.

[3] Higgins, C. H., "Report of Veterinary Director General for 1905." Dept. of Agr., Canada, Ottawa, 1906.

disease has been reported in many other places within the last few years. It thus seems certain that the disease is widespread throughout the United States and Canada and in the future must be reckoned with by American poultrymen.

Tuberculosis may exist extensively among fowls, especially in large flocks, and yet not kill enough birds to attract attention to it. Reports show that farmers often lose one or two birds a year from what appears to be tuberculosis. In many places the loss seems to be gradually increasing. The existence of the disease in the flock fails to attract the attention of the owner because the losses are so evenly distributed throughout the year. In other cases the disease appears to be more virulent and to cause very serious losses. Moore and Ward report a flock of 1400 birds from which 250 had died during the first year. Another man lost 300 birds out of a flock of 1460. Microscopic examination proved that these were dying of tuberculosis.

Tuberculosis is confined chiefly to adult or nearly adult fowls. Only very rarely, if ever, is it found in growing chicks. Further it is much more common in fowls than in other kinds of poultry. Two cases in wild geese were reported at the Ontario Agricultural College. Avian tuberculosis is said to be found in turkeys, pheasants, and especially in pigeons. Cage birds are particularly susceptible to this disease.

Etiology. — Tuberculosis is caused by a minute germ, the *Bacillus tuberculosis* of birds. These bacteria gain entrance to certain portions of the body and there multiply in vast numbers, causing the formation of small nodules or tubercles. The disease is highly contagious and is spread through the flock by the contact of healthy birds with the diseased ones, or with their discharges.

The relation of avian tuberculosis to that of man and other animals has attracted a great deal of attention. It is a subject of very great importance to the poultryman, not only on account of his flock but also on account of its relation to the health of himself and his family. The bacillus associated with avian tuberculosis presents certain morphological

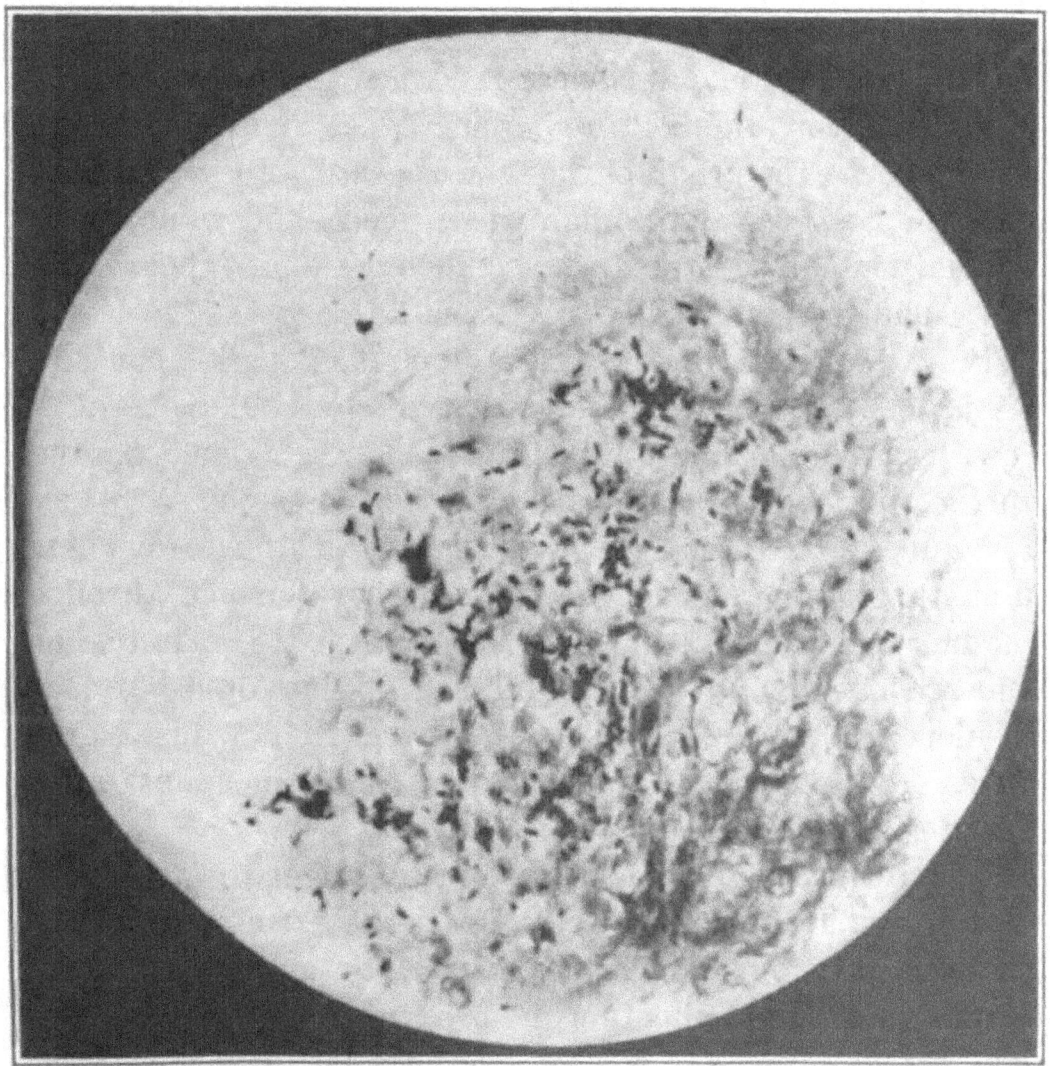

Fig. 11. — Section of the lung of a hen showing tubercle bacilli. (After Himmelberger.)

and physiological characteristics which are different from the organism associated with this disease in man. Likewise the bacillus of human tuberculosis is differentiated in certain marked features from that found in cattle. On the whole the difference between the avian and the other two types is much greater than that between the human and the bovine.

It has frequently been held that the avian bacillus is a distinct species. The chief reason for this is that it often fails to produce disease when inoculated into mammals and because the mammalian type will not always infect birds. At the present time the view is rather generally accepted that the avian, human and bovine types simply represent three varieties or strains of the same species. A large amount of work has been done upon this subject, and while under ordinary conditions the avian bacillus does not infect mammals, under certain conditions it will do so.

Bang,[1] who has done a great amount of work upon this subject, found that mammalian bacilli by passage through fowls can be so changed as to behave like the avian type, and further that bovine bacilli after having lost their virulence for guinea pigs through repeated passage through fowls are able to regain the original virulence by passage through mammals. Of eighteen different strains of mammalian tubercle bacilli used Bang found that twelve could be made virulent for fowls. He states, however, that in his experience mammalian bacilli were never found in spontaneous avian tuberculosis.

The avian tubercle bacilli are very virulent to most birds and especially to domesticated species. Artificial infection succeeds best by direct inoculation into a vein, while intraperitoneal and subcutaneous injections are apt to yield less certain results. By feeding either cultures or fresh material from tuberculous birds the disease is readily trans-

---

[1] Bang, Oluf, "Die Tuberculose der Geflügels in ihren Beziehungen zur Tuberculose der Säugethieren." Trans. IX *Intern., Vet. Cong.*, Vol. 1, 1909.

———"Geflügeltuberculose und Säugetiertuberculose." *Centralb. f. Bakt. Paras. u. Infekt.*, Bd. XLVI, 1908.

mitted. The fresh material has usually proven to be the most virulent. Van Es and Schalk [1] report that of 12 English sparrows each fed one meal of chopped tuberculous chicken liver all died in from 73 to 202 days with generalized tuberculosis. It is quite probable that the English sparrow often serves to infect domestic fowls.

Koch and Rabinowitsch [2] state that while fowls are easily infected with avian tuberculosis by feeding, it is very difficult to infect them with the mammalian strains in the same way. On the other hand, some birds, especially cage birds, are very readily infected in various ways. Parrots, in particular, are susceptible not only to avian tuberculosis, but also to mammalian and human tuberculosis. Also canary birds, sparrows, and various birds of prey were proven to be susceptible to both avian and mammalian tuberculosis. In these respects such birds differ materially from the domestic fowls.

On the other hand their later researches have made it apparent that a large number of mammals are susceptible to avian tuberculosis. These include not only the small laboratory animals as rabbits, mice and guinea pigs, but also cattle, hogs, horses, goats, and donkeys. Also avian tubercle bacilli have been found in cases of human tuberculosis.

Himmelberger [3] reports experiments in which it was possible to infect a calf by feeding it the macerated organs of a tuberculous hen. This result is of considerable interest in view of the question of the relation of the avian tubercle bacillus to the causative factor in Johne's disease of cattle. Johne's disease presents many of the symptoms of tuber-

[1] Van Es, L., and Schalk, A. F., "Avian Tuberculosis." North Dakota Agr. Expt. Stat. Bul. 108, pp. 1-94, 1914.

[2] Koch, R., and Rabinowitsch, L., "Die Tuberculose der Vögel und ihre Beziehungen an Säugetiertuberculose." Arbeiten a. d. Kaiserl. Gesundheitsamte, 1904.

[3] Himmelberger, L. R., Centralb. f. Bakt. etc. Abt. 1, Bd. 73, pp. 1-11, 1914.

culosis in cattle, yet usually such cattle do not react to the ordinary tuberculin test. However, it has been found that in a considerable number of cases of this disease the animals will react if tested with a tuberculin made from the avian bacillus. The majority of experimenters have reported negative results in their attempts to infect cattle with the avian organism. The question is one which must await further evidence before definite conclusions can be drawn.

On the basis of such experiments and observation it appears that the difference between avian and mammalian tuberculosis has developed because the bacilli have grown for a long time under different conditions. They are not so different, however, but that each may grow in the environment best suited to the other.

It thus appears that while fowls are not very likely to contract tuberculosis from domestic animals or from man, yet fowls that have the diseases are a serious menace to the other animals on the farm as well as to the poultryman and his family. (Cf. further on this point p. 128 below.)

Diagnosis. — Tuberculosis in mankind is so serious a disease chiefly because it is so difficult to recognize it in its earliest stages. The same is true with the disease in fowls. There are positively no external symptoms by which the disease can be recognized in fowls before the advanced stages. Some of the outward symptoms that may serve to arouse suspicion are: steadily advancing emaciation; anæmia, shown by pallor of comb, wattles and the skin about the head; general weakness; lameness; ruffling of the feathers, and in many cases diarrhea. These combined with a bright eye and a ravenous appetite are some of the symptoms most frequently found. None of them is specific, however, and final diagnosis must be based on other findings. Emaciation is one of the best symptoms and in the last stages of the disease becomes very marked. Pernot cites the case of a Plymouth Rock hen weighing 4 pounds that was reduced to 22 ounces. The emaciation is very marked in the muscles covering the breastbone.

Lameness is another symptom often shown in the later stages of the disease. This is caused by tuberculosis of the joints, as has been proven in many cases. Such cases are often called "rheumatism" by poultrymen. Tuberculosis may also form tumors or ulcers or various outgrowths on the head and limbs of birds. Such forms of the disease are comparatively rare in poultry, however. Parrots are particularly affected with these external tubercles.

None of these symptoms, however, is more than an indication of the possible presence of the disease.

Post-mortem find-

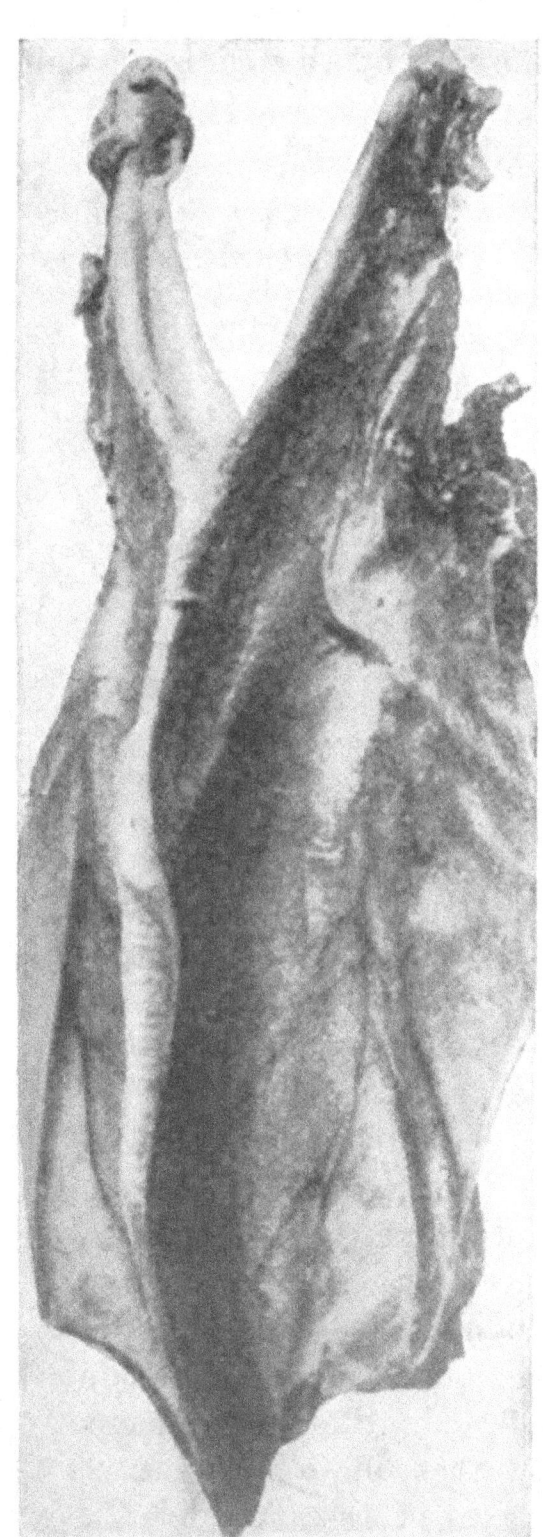

Fig. 12.—Breastbone of a fowl showing excessive emaciation in tuberculosis. (After Wrad.)

ings give much more certain evidence of the existence of this disease. The tubercle is the unit of all tuberculous lesions.

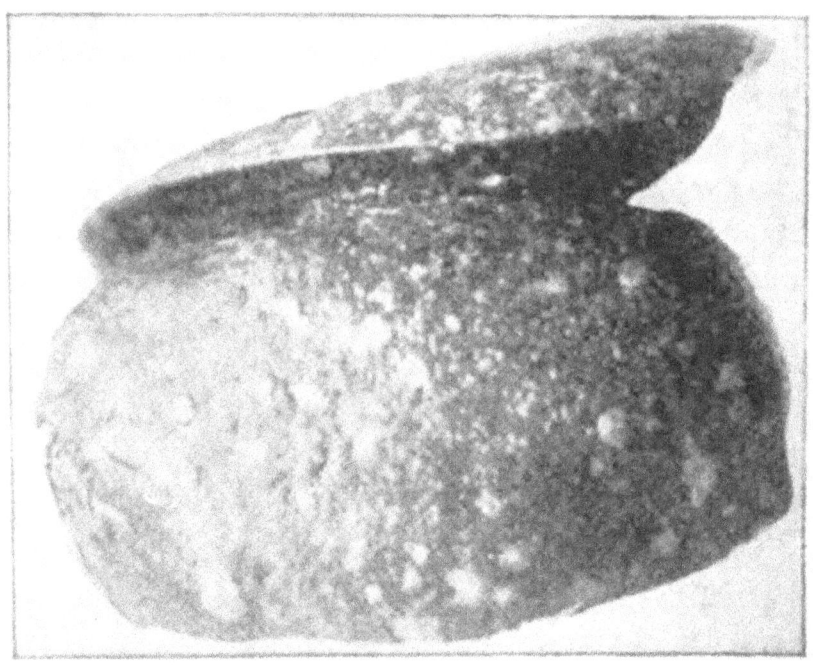

FIG. 13. — Liver of fowl affected with tuberculosis. (After Ward.)

The tubercles in avian tuberculosis are not essentially different from those found in mammalian forms of the disease. These tubercles appear as small raised nodules filled with a cheesy substance.

In birds the organs most affected are the liver, spleen and intestinal tract. In some instances nearly every organ, including kidneys, ovaries, lungs, bones, muscles and skin, is affected. Statistics collected show that in from 90 to 99 per cent of cases the liver shows tubercular lesions (Fig. 13). In from 85 to 90 per cent the spleen is affected

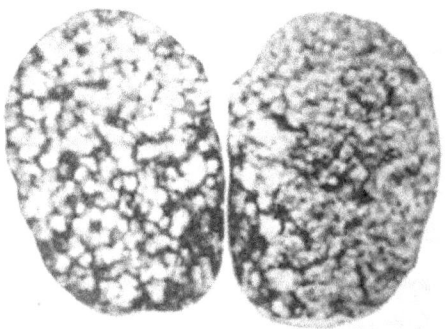

FIG. 14. — Spleen from tuberculous fowl cut through the middle. (After Koch and Rabinowitsch.)

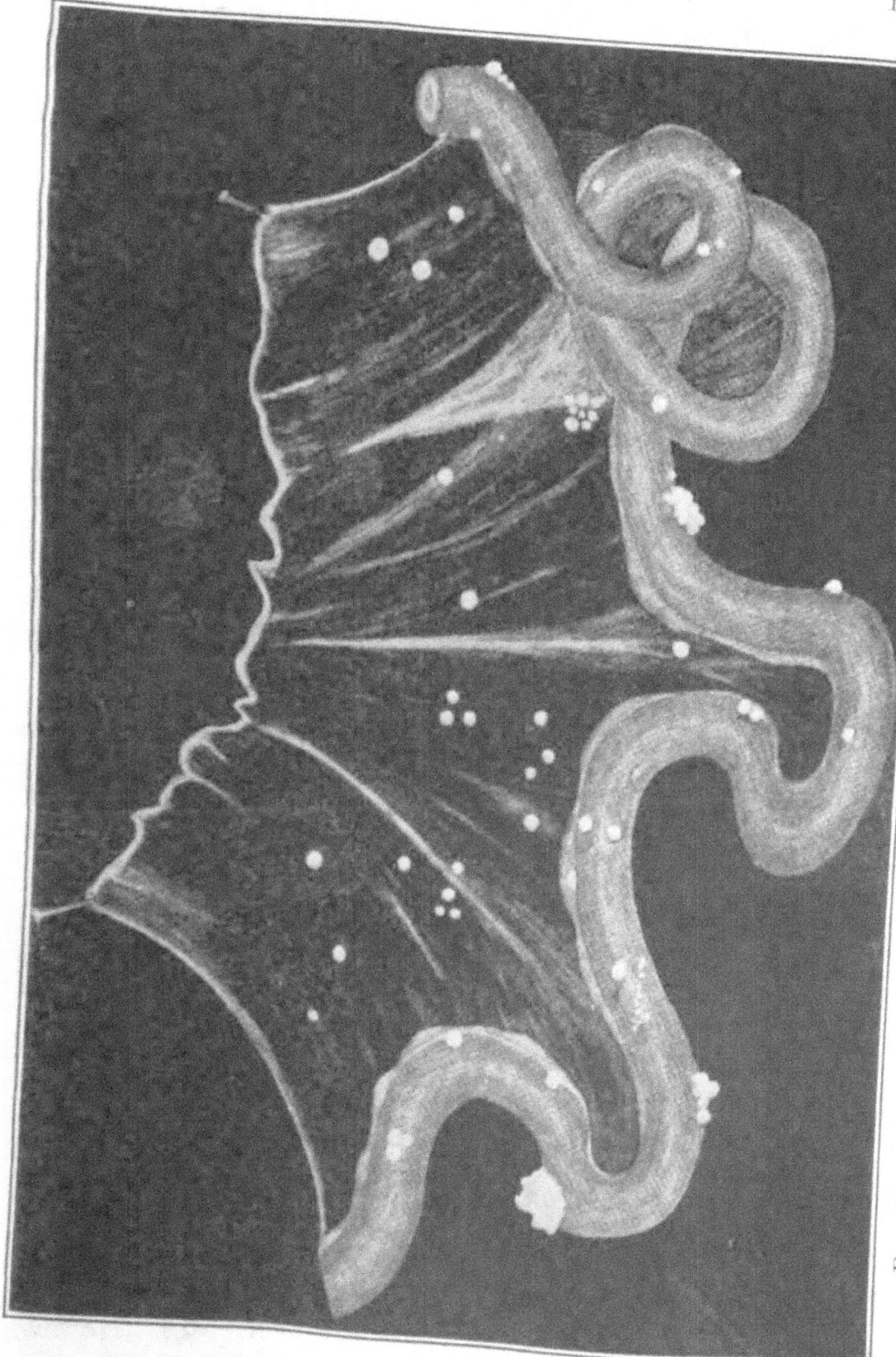

Fig. 15.—Tuberculosis of the intestines and mesenteries of a fowl. (After Van Es and Schalk.)

Fig. 16. — Intestine and mesenteries of a fowl affected with tuberculosis. (After Ward.)

(Fig. 14) and in from 50 to 60 per cent of the cases tubercles are found in the intestines and mesenteries (Figs. 15 and 16).

Thus the liver is affected in nearly every case. However,

as has been pointed out many times in these pages, a spotted condition of the liver is no sure sign of tuberculosis. Most of the other liver diseases of fowls cause a simple blotching of the tissue in which the center of each spot is usually depressed or at least only slightly raised (cf. Fig. 9, p. 95). In tuberculosis the liver is covered with numerous raised nodules varying greatly in number and size as shown in Fig. 13. A section of the liver shows these nodules or tubercles distributed throughout the tissue.

Still more conclusive evidence is found if the spleen is covered with these same kind of nodules. The spleen in health is a small rounded purplish organ about ½ inch in diameter. It lies just above the liver in the region of the gall bladder. (Cf. Fig. 7). In cases of tuberculosis it is very frequently greatly enlarged and is studded throughout with the yellowish-white tubercles as shown in Fig. 14.

The lungs are *rarely* affected and then usually by the infection spreading from the liver on to the adjoining lung tissue. All this agrees with the fact previously stated that fowls are most easily infected through the digestive tract.

If the post-mortem findings agree in essentials with those given in the preceding paragraphs we may be practically certain that we are dealing with tuberculosis. It should not be forgotten, however, that the pathologist would not be willing to pronounce the disease tuberculosis until he had taken a small particle of the cheesy material and after staining this in a particular way had demonstrated by microscopical examination that the tubercule bacilli were present.

Some recent experiments hold out the hope that means will be found for the accurate diagnosis of this disease by means of some of the biological and serological tests. Van Es[1] and Van Es and Schalk[2] have carried out careful experi-

---

[1] Van Es, L., *Zeitschr. f. Infektionskrank. u. Hyg. Haust.* Bd. 14, pp. 271–296, 1913.   [2] *Loc. cit.*

ments with tuberculin prepared from the avian bacilli. These authors injected this tuberculin within the layers of the skin (intracutaneous) in the comb and wattles. The results indicate that this method is reasonably satisfactory. Summarizing the results of their tests on 601 birds, these authors found that 98 per cent of the birds showing tuberculous lesions reacted to the test. Only 8 per cent of the non-reacting birds showed lesions. Numerous previous investigators had obtained negative results with the subcutaneous, ophthalmic and cutaneous method of application. It appears that for birds the intracutaneous method is the only one to be considered of value.

Technique of making the injection is very important according to Van Es and Schalk. Care must be taken that the injection is neither too deep nor too superficial. A small caliber syringe with a very fine needle (No. 26 or 27) is used.

Fig. 17.—Syringe used in the tuberculin test for chickens. (After Van Es and Schalk.)

Owing to the denseness of the tissues, especially the comb, it is often difficult to inject even a small quantity, and considerable pressure is required to force the tuberculin into the tissues. The amount injected into each bird was not carefully measured but varied between $\frac{1}{30}$ and $\frac{1}{20}$ of a cubic centimeter of a 50 per cent avian tuberculin. This tuberculin was prepared in the usual way.

Owing to its looser texture the wattle proved to be the best place for the injection. The results of the tests were recorded 24, 48 and 72 hours after the injection. A positive reaction is indicated in a typical case by a large swelling

Fig. 18. — Head of chicken showing positive tuberculin reaction of comb and right wattle. (After Van Es and Schalk.)

about the point of injection. In the wattle this organ often becomes two to three times its original thickness.

Agglutination and complement fixation tests have also been used to diagnose this disease.[1] The tests so far re-

[1] Himmelberger, L. G., *loc. cit.*

ported, while encouraging, are too few to allow of definite conclusions.

Methods of Contagion. — The spread of tuberculosis from fowl to fowl takes place only when the living bacteria are transferred from the diseased to the healthy birds. From the fact that tuberculosis lesions are most commonly found in the internal organs of the digestive system we may conclude that the bacteria usually enter the body along with the food. Examination of the tubercles situated along the intestine shows that in many cases these communicate directly with the interior of the digestive tract. These are constantly emptying enormous numbers of bacteria which are carried to the outside by the feces of the bird. Without doubt the droppings of tuberculous fowls are the most important factor in the spread of this disease. This is especially true when in addition the birds are fed upon ground which is partly covered with these droppings. Besides, the infectious material may very easily be carried by the feet and thus mixed with the food.

Ward states that there is no evidence to indicate that tuberculosis is spread through the egg. He cites in support of this first the fact that badly diseased birds do not lay, and second the absence of tuberculosis among young stock. Other authors, however, have collected statistics which indicate that even birds badly diseased with tuberculosis may continue to lay quite steadily.

Koch and Rabinowitsch also make the following statement (p. 431): "The possibility of the congenital origin of tuberculosis of fowls through the infection of the fertilized egg with bird tuberculosis is shown by our results. It is also demonstrated by our inoculation experiments on eggs." Further they have given experimental proof of the transfer of the bacteria of mammalian tuberculosis from the *inoculated egg* to the chick.

Lowenstein [1] states that avian tuberculosis occurs more frequently in man than is usually supposed and that it may be due to eating eggs from tuberculosis hens. Artificially infected eggs still contain living organisms after having been soft boiled.

In this connection it is of interest to mention a case of the apparent transfer of fowl tuberculosis to man. In the *Medical Record* (Vol. 31, 1887) there is recorded a case of human tuberculosis in France which apparently came from eating tuberculous fowls which "were cooked very little before being eaten." The case occurred "in a little hamlet of 10 cottages isolated in the midst of a large forest." No other source of infection could be discovered.

Treatment. — Fowl tuberculosis when it reaches the stage at which it can be diagnosed cannot be cured under our present knowledge. Treatment of individual cases should not be attempted. Salmon [2] says: "When the disease is discovered the effort should be to eradicate it at once by killing off the whole flock and thoroughly disinfecting all the houses and runs.

"As the great majority of the birds will probably be more or less affected, the chances are that any which are saved will have diseased livers and intestines, from which the bacilli will escape and keep up the infection of the flock and the runs. The danger of this is so great that no attempt should be made to keep any of the fowls that have been exposed to the contagion, no matter how valuable they may be. The bodies of the birds which have died or are killed, as well as all the accumulated manure, sweepings, and scrapings of the poultry houses, should be completely destroyed by fire."

---

[1] Lowenstein, E., "Ueber das Vorkommen von Geflügeltuberculosis beim Menschen." *Wiener Klin. Wochenschrift*, Bd. 26, pp. 785–787, 1913.

[2] Salmon, D. E., "Important Poultry Diseases." U. S. Dept. of Agr. Farmers' Bul. 530, 1913.

The above recommendations, while drastic, will probably prove the wisest in the long run. There are, however, sometimes mitigating circumstances under which it would not be advisable to do this. If it is known that the disease has recently been introduced or that it is not very widely spread through the flock all of the old stock should be removed and killed for table purposes, providing their condition permits. Van Es and Schalk have shown that it is the older birds that furnish the higher percentage of actual disease. Their autopsy records show the following:

Age of Bird	Per Cent of Infected Birds
1	3.33
2	24.35
3	86.44
4	85.71

Van Es and Schalk recommend that "After the elimination of the older birds the remainder of the flock may be tuberculin tested and all fowls reacting typically or doubtfully should share the same fate as the older birds."

After the diseased birds have been disposed of the houses, runs, eating and drinking utensils should be thoroughly cleaned and disinfected. Everything loose should be burned. The disinfecting so far as possible should be done by boiling and by sunlight. Most of the common disinfectants cannot be relied upon to kill the tubercle bacilli. Heat and sunlight are very effective wherever they can be applied directly. The runs should be cultivated and the houses should be open to the sunshine and fresh air at all times.

Van Es and Schalk report a flock which at the beginning of 1913 had 249 chickens. Of these 43.37 per cent were found to be tuberculous by the tuberculin test and autopsy. All reacting and undesirable birds were eliminated, leaving 56 non-reacting fowls to which 47 were added by purchase. One year later a similar test of this flock, which had again increased to 249 birds, showed only 2.41 per cent tuberculous. It would seem that the measures reported might hold out hope that in slightly infected flocks the disease may be eliminated.

In the majority of cases, however, the cost of administering the tuberculin, which would have to be done by an experienced veterinarian, would be more than the birds were worth. In such cases it will probably be best to kill off the old stock and after thoroughly cleaning and disinfecting start new with stock known to be healthy.

If it is particularly desired to maintain the same strain of birds it might be done by adopting a method similar to that proposed by Bang for new herd building in the case of tuberculous cattle. Directions for doing this are given by Morse[1] as follows:

"Secure new or thoroughly disinfected ground, keeping it absolutely free from contact with the ground used by the infected flock. Erect new houses on this ground. Collect the eggs from the infected birds and wash them in 95 per cent alcohol or in a 4 per cent solution of some good coal tar disinfectant. Incubate these disinfected eggs in new incubators. When hatched, remove the chicks to new brooder houses on the new ground. These growing chicks should be cared for by new men, that is to say, either different men from those that care for the old flock, or if you are compelled to use the same men they should disinfect their hands and shoes and put on fresh overalls before handling the new stock. Have different feed bins and different pails for distributing it. As soon as you have built up a clean flock

[1] *Reliable Poultry Journal*, 1910.

destroy the old and disinfect the ground occupied by them by the method outlined above."

This method is, no doubt, excellent in theory and if carried out with complete and *never-failing* attention to details might work. It is doubtful, however, whether in actual practice a poultryman would ever be able to carry it through successfully or profitably.

CHAPTER X

Internal Parasites

Fowls are often seriously infested with internal parasites. The most important of these are various worms living in the alimentary canal. In popular usage these are spoken of simply as "worms." Various other internal parasites, as the gape worm, the air-sac mite, etc., are described in other sections of this book. In the main the present discussion will be confined to intestinal worms.

Few flocks of poultry or indeed few birds could be found which are free from intestinal worms. Worms of one kind or another are found in the intestinal tracts of practically all fowls. Under ordinary conditions these parasites do no very serious harm. Undoubtedly the bird would be better off without them but they are not serious enough to be worth troubling about. Under certain conditions, however, these parasites may multiply to such an extent that they become a serious menace to the flock. There are several cases on record in recent years where epidemics of worms have put whole poultry plants out of business.

Worms are spread from bird to bird usually through the excrement. The worms or their eggs are expelled by one bird and are picked up along with food and grit by another. Some forms are taken in with the drinking water, especially where fowls are allowed to drink from stagnant pools. Still other forms, like the tape worms, require an intermediate host such as an angleworm, snail, or insect.

Diagnosis of Worms in General. — Accurate diagnosis of worms in the intestines can be made only by finding the worms in the droppings of the fowls. Fowls affected with worms to any great extent frequently show the general symptoms of dullness and depression. Birds that are suspected of being affected with worms should be shut up in a coop and given a dose of some vermifuge or a purgative dose of Epsom salts. If careful observation of the droppings is made at frequent intervals the worms, if present, can usually be detected in this way. This is not, however, an infallible test.

If there is any reason to suspect that worms are present in the flock one or two birds showing the most advanced symptoms should be killed and examined. The entire digestive tract should be opened and the contents carefully examined. The intestines should be washed out in a gentle stream of water and their walls examined after immersing in a pan of water. If tape worms or other parasites which are attached to wall are present these can be seen readily under water. In case there is any doubt a competent veterinarian should be consulted, or a bird may be sent to the Zoölogical Division of the Bureau of Animal Industry of the United States Department of Agriculture, Washington, D.C. In this way the worms will be identified and any specific remedies will be recommended.

The principal parasitic worms which affect the digestive tract of fowls may be grouped into three classes as follows: Tape worms, round worms and flukes.

Tape Worms

Tape worms have long been known to infest domestic poultry. Occasionally serious outbreaks of the tape worm disease occur in various parts of the country. These out-

Internal Parasites

breaks are usually confined to comparatively small areas and are perhaps more common in the southern states.

Etiology. — The tape worms of poultry, like those which infest man and the domestic animals, are long, *flat, segmented* worms (Fig. 19). The anterior end of the animal possesses a number of hooks or suckers by which it attaches itself to the walls of the intestine. Back of this head the entire animal consists of a long series of segments or proglottids. The segments nearest the head are the smallest and it is at this region that new segments are constantly being formed. The farther from the head they get the larger the segments become. Towards the posterior end of the worm the segments develop sexual organs and later become filled with eggs. As soon as the eggs are fertilized and mature the segment containing them drops off and passes to the exterior with the feces of the host. Each segment of this kind contains thousands of eggs.

If these eggs are to develop farther they must be swallowed by some intermediate host (as a worm, snail or insect). The egg then hatches into a 6-hooked embryo which bores its way from the intestine into the body cavity of the inter-

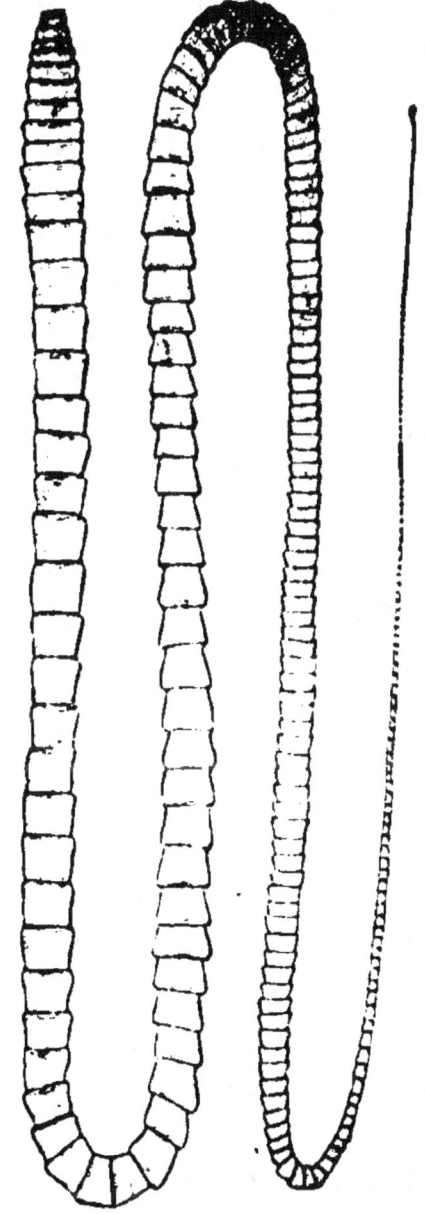

FIG. 19. — *Drepanidotænia infundibuliformis*, a tape worm of the fowl. (After Stiles.)

mediate host. It here develops into a larval form known as a *cysticercoid*. When the intermediate host (worm, snail, etc.) is eaten by a chicken this larva continues its development and forms an adult tape worm. Thus there are two stages in the life cycle of a tape worm: that in the adult host and that in the intermediate host. Each species of tape worm, of which there are a great many, has its particular host, both intermediate and final.

According to Stiles [1] there were up to 1896, 33 species of tape worms recorded for poultry. Of these 11 are recorded as occurring in chickens (*Gallus*). The complete life history is known for only a few of these. Since that time several other species have been described.[2]

Regarding the tape worms of chickens, Stiles (*loc. cit.*) says, p. 13: "(They) are *known* to become infected with one tape worm through eating slugs (*Limax*). They are *supposed* to become infected with a second through eating snails (*Helix*); by a third through eating flies and by a fourth through eating earth worms."

There seems but little need to give a description of the different species of tape worms found in chickens. The characters by which they are distinguished from each other are too minute and involved to be of use to the poultryman or farmer. If any one is having trouble with tape worms in poultry the best thing to do is to send a portion of the intestine containing the worms to Washington as directed above. The correct identification of the species and the corresponding knowledge of its life history will often suggest a specific means of control.

---

[1] Stiles, C. W., "The Tapeworms of Poultry." U. S. Dept. of Agr., Bur. of Anim. Ind., Bul. 12, pp. 1–80, 1896.

[2] See Ransom, B. H., "The Tapeworms of American Chickens and Turkeys." U. S. Dept. of Agr., Bur. of Anim. Indus., Ann. Rept., 1904, pp. 268–285.

Nodular Tæniasis

Stiles says, p. 15: "At least one species of tape worm (*Davainea tetragona*) causes a serious nodular disease of the intestine of chickens which upon superficial examination may be easily mistaken for tuberculosis." Moore [1] says:

"Tuberculosis is the only known disease for which this affection is liable to be mistaken, and it is of much importance that the two diseases should not be confounded. The diagnosis has not in my experience been difficult, as in every case the attached tape worms were readily detected upon a close examination of the intestinal contents, or of the mucous membrane of the infected portion of the intestine. However, the worms are quite small and could easily be overlooked in a hurried or cursory examination. In case of doubt, if the affected intestine is opened and the mucous surface washed carefully in a gentle stream of water, the small worms will be observed hanging to the mucous membrane. This discovery, in the absence of lesions in the liver or other organs, would warrant the diagnosis of the tape worm disease."

Diagnosis. — The symptoms of tape worm disease are not specific. The general symptoms are similar to those of other worms (cf. p. 134). Regarding the symptoms of tape worms Zürn [2] says:

"If numerous tape worms are present in the intestine of young or old fowls a more or less extensive intestinal catarrh develops, corresponding to the greater or less number of parasites present.

"The intestinal catarrh shows itself, especially in chickens

---

[1] Moore, V. A., "A Nodular Tæniasis in Fowls." U. S. Dept. of Agric., Bur. Anim. Ind., Circ. No. 3, p. 4, 1895.

[2] Zürn, F. A., "Die Krankheiten des Hausgeflügels." Weimar, 1882.

and geese, as follows: The sick animals become emaciated, although the appetite is not especially disturbed. At times the appetite is even increased. The droppings are thin, contain considerable yellow slime, and are passed in small quantities but at short intervals. The poultry raiser must direct his attention to these thin, slimy, and often bloody droppings, for if any treatment against the tape worm is to be undertaken, this must be done as early as possible. In observing the droppings it should be noticed whether tape worm segments or eggs are present. The eggs can be seen, of course, only with the microscope.

"After a time other symptoms develop. The sick animals become dull and listless, remain apart from the rest of the flock — the feathers are ruffled and the wings droop, the appetite is lost and the birds allow themselves to be easily caught. Although it was stated that in the beginning of the trouble the appetite is not disturbed, the sick animals develop an intense thirst for cold water. When it rains they run under the eaves in order to catch water, and in winter are eager for ice water."

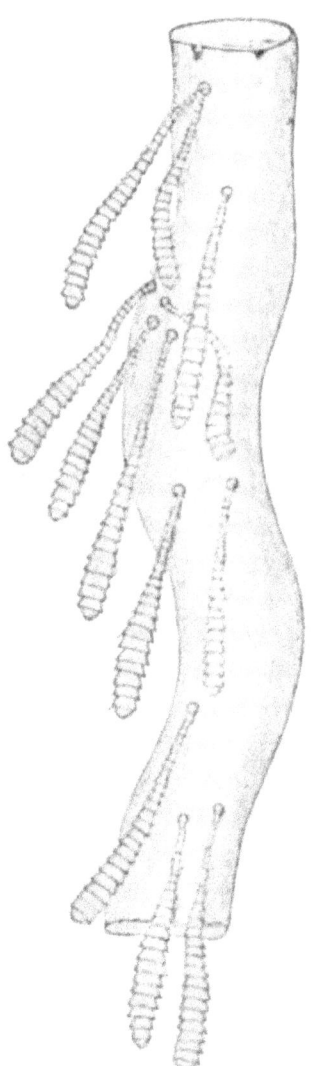

FIG. 20. — Intestine of a fowl turned wrong side out to show tape worms in nodular tæniasis. (After Pearson and Warren.)

Since the examination of the feces for tape worm segments is rather unsatisfactory for the farmer or poultryman, Stiles says that "The best method for the farmer to follow is to kill one of the sick chickens when he suspects tape worms and to cut out the intestine. He should then open the intestinal tract from gizzard to anus in a bowl of warm water, and look for the parasites" (cf. Fig. 20). Finding the worms in the alimentary canal is the only certain diagnosis of the disease.

Treatment. — The chief drugs used for tape worms in fowls are: Extract of male fern, turpentine, areca nut, powdered kamala, pumpkin seed, pomegranate root bark and Epsom salts. The following extract from Salmon gives the principal methods of treatment and the doses: "One of the best methods of treating tape worms in fowls is to mix in the feed a teaspoonful of powdered pomegranate root bark for every 50 head of birds. In treating a few birds at a time it is well to follow this medicine with a purgative dose of castor oil (2 or 3 teaspoonsful). According to Zürn, powdered areca nut is the best tape worm remedy for fowls, but he states that turkeys are unfavorably affected by it. It may be given in doses of 30 to 45 grains mixed with butter and made into pills. Male fern is also a very effectual remedy and may be used in the form of powder (dose 30 grains to 1 dram) or of liquid extract (dose 15 to 30 drops). It should be given in the morning and evening, before feeding. Oil of turpentine is an excellent remedy for all worms which inhabit the digestive canal. It may be given in the dose of 1 to 3 teaspoonsful, and is best administered by forcing it through a small flexible catheter that has been oiled and passed through the mouth and esophagus to the crop. This medicine is less severe in its effects if diluted with an equal bulk of olive oil, but, if it fails to destroy the parasites when so diluted, it may be given pure."

A method of administering medicine such as turpentine by depositing it directly in the crop has been proposed by Gage and Opperman.[1] This method can be advantageously

[1] Gage, G. E., and Opperman, C. L., "A Tapeworm Disease of Fowls." Maryland Agr. Expt. Stat. Bul. 139, pp. 73–85, 1909.

used with many other liquid remedies, and should be adopted in all cases where it is important to have the full dose in the stomach in a short time. It does away with the uncertainty attending the giving of medicine in the feed or drinking water, and with a little practice is more expeditious than making

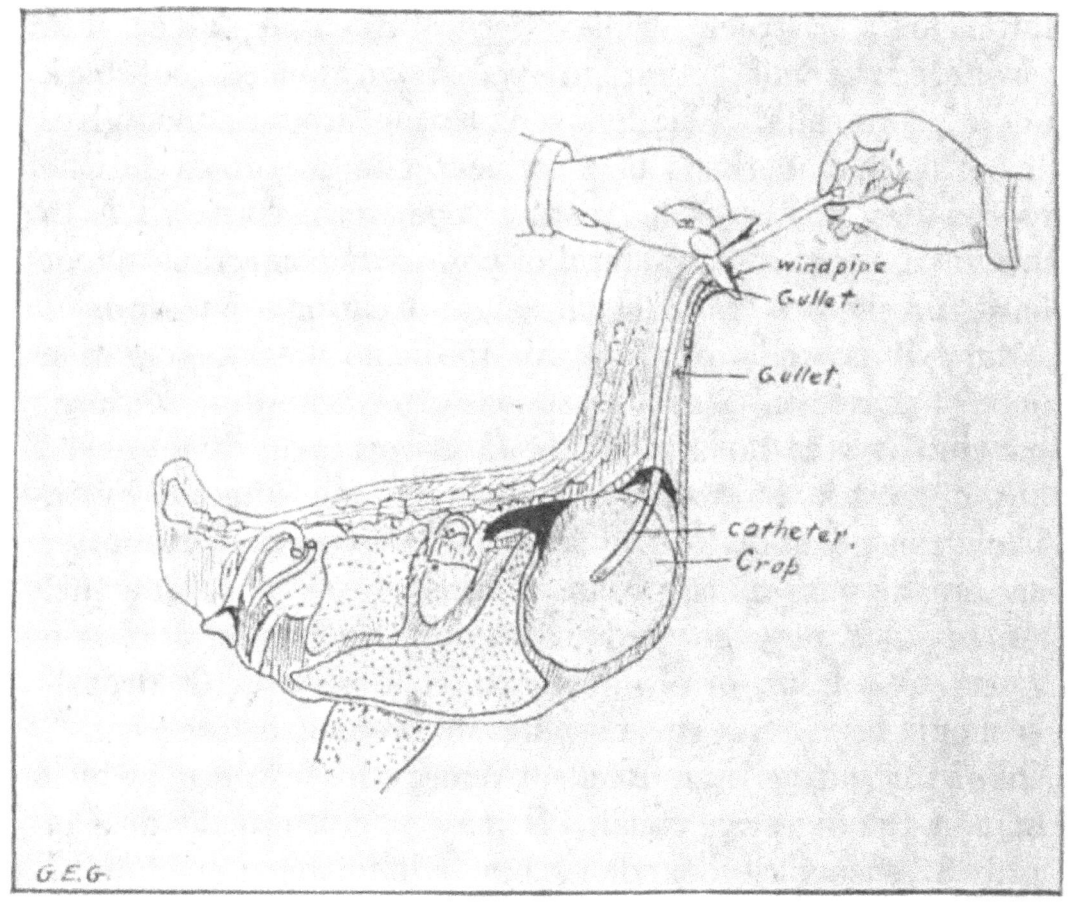

Fig. 21. — Sketch showing method of introducing turpentine directly into crop. (After Gage and Opperman.)

and giving pills. The open end of the catheter may be inserted into a rubber bulb having one opening. Just sufficient air should be expelled from the bulb, so that the dose of medicine will be sucked up without being followed by much air. The bird's head is then brought in a line with the neck, which is extended, the catheter is passed carefully to the crop (Fig. 21), when a slight pressure on the bulb forces out the medicine, and the instrument is withdrawn. The operator should be sure that he avoids the trachea.

Gage and Opperman have found Epsom salts and turpentine a very effective remedy for Nodular Tæniasis. After careful consideration of the data they conclude that "40 to 50 grains of Epsom salts is sufficient for an adult fowl in order to clean out the intestinal tract so that the birds may take food. Then the turpentine should be introduced" as directed above. For younger birds the dose of salts should be proportionately less. In fowls from 6 months to 2 years old the salts are best given by dissolving in water and giving each fowl this liquid. For younger chicks the salts may be dissolved in warm water and used to moisten the mash or feed.

Prevention. — One of the most important measures against all parasitic infestations of the digestive tract is to move the fowls upon fresh ground every two or three years. This should be done in all cases where such parasites are frequently observed in the intestines of the birds. Another practical measure, which may be adopted at the same time, is to remove the excrement daily from the houses and destroy any parasites or their eggs which may be in it, by mixing it with quick lime or saturating it with a 10 per cent solution of sulphuric acid. The acid is cheap, but requires that great care be taken in diluting it, owing to danger of its splashing upon the clothing and flesh and causing severe burns. It should always be poured slowly into the water used for dilution, but on no account should water be poured into the acid as it will cause explosions and splashing.

When treating diseased birds these should always be isolated and confined, and their droppings should either be burned or treated with lime or sulphuric acid as just recommended. Without these hygienic measures, medical treatment can only be partially successful.

Stiles says: "An extermination of slugs will insure immunity against *Davainea proglottina*, but no precise directions can be given to prevent chickens from becoming infected with other tape worms until the life history of these parasites is better understood. It will be well, however, to keep the chickens housed in the morning until the sun is well up and the ground is dry, for they will thus be less likely to meet with the supposable intermediate hosts of other worms."

Round Worms

Round worms can be found in the intestine and especially the ceca of almost any fowl. They are much more common than the tape worms. Normally the round worms cause

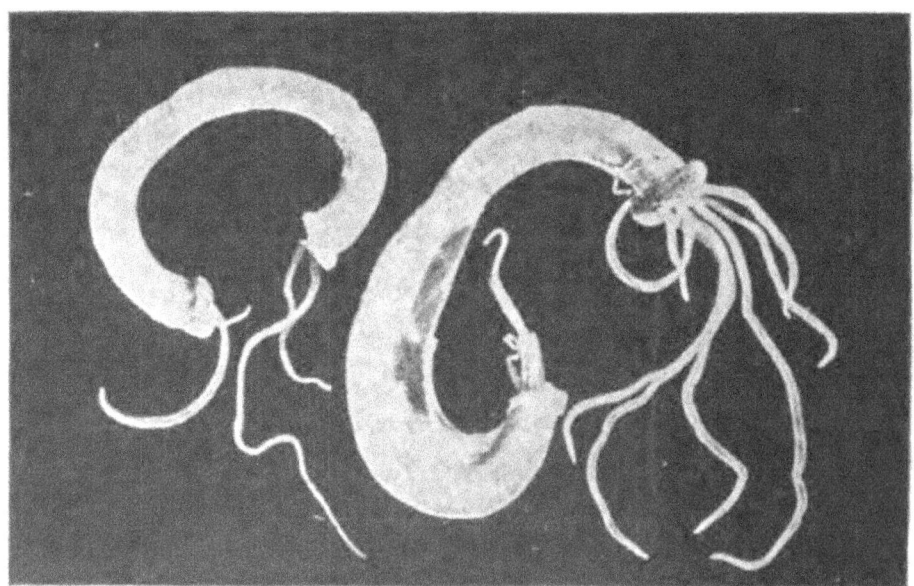

Fig. 22. Worms protruding from a section of the intestine of a fowl. (After Bradshaw.)

no serious trouble to fowls. Under certain conditions, however, they may become so numerous as to be a serious menace to the flock. At such times they have a decided effect on the digestion; the irritation often causes diarrhea. When in large numbers, they sometimes become rolled and matted into a ball which may cause complete stoppage of the intestine.

The round worms are white in color and vary in length from ⅛ inch to 5 inches. The head end is sharp pointed, while the tail end is more blunt. Round worms are seldom passed in the feces unless present in very large numbers. When a worm is passed it soon dies in the droppings or is eaten by another fowl.

Dispharagus spiralis, a small worm about ⅛ inch in length, is often found in the esophagus and occasionally in the crop or intestine.

Dispharagus nasutus, about ¼ inch long, occurs in the walls of the gizzard of fowls. It sometimes becomes so numerous as to cause serious loss.

Another nematode, *Cheilosperura hamulosa*, parasite in the gizzard of the chicken has recently been recorded in this country by Ransom.[1] Specimens have been found from the District of Columbia as far west as Kansas.

Two other nematodes, *Trichosoma strunosum* and *Gingylonema ingluvicola*, have been found in the pharynx and esophagus of chickens.[2]

Heterakis perspicillum, from 1½ to 3 inches long, is very common in the intestines of fowls. They sometimes become very numerous and may become rolled into rather large balls which obstruct the passage of the food.

Scott[3] has found that this nematode may be transmitted to young chicks through an earthworm (probably *Helodrilas*

---

[1] Ransom, B. H., "The Occurrence of *Cheilosperura hamulosa* in the United States." *Science*, N. S., Vol. 35, p. 555, 1912.

[2] Crurea, J., *Zeitschr. Infekt. u. Hyg. d. Haust.* Bd. 15, pp. 49–60, 1914.

[3] Scott, J. W., "A New Means of Transmitting the Fowl Nematode, *Heterakis perspicillum*." *Science*, N. S., Vol. 38, pp. 672–673, 1913.

parvus) found in horse manure. Whether the worm is an intermediate host or whether the nematode eggs simply cling to the surface of the worm has not been determined.

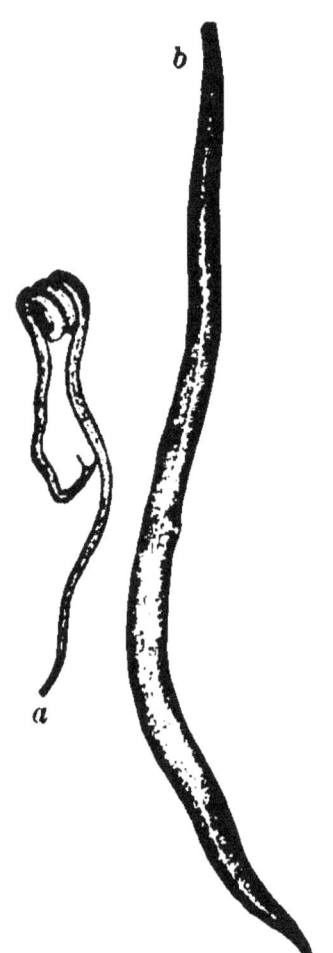

Fig. 23. — *Heterakis perspicillum.* a, male. b, female. × 2. (After Salmon.)

Several other species of the genus *Heterakis* also infest fowls and other poultry.

At least one species of ascaris (*Ascaris inflexa*) is found in the fowl. This is a round worm white or yellowish-white in color and from one to two inches in length. If they occur in sufficient numbers they produce considerable irritation in the digestive tract. Infested birds appear unthrifty, lack appetite and become emaciated.

Occasionally this or other round worms may pass from the cloaca into the oviduct or egg tube. In this way they may be incorporated in the albumen of an egg as it is formed in the oviduct.

Diagnosis. — The symptoms of round worms are similar to those of all worms (cf. p. 134). There is evidence of indigestion. The comb becomes pale and there may be diarrhea.

Treatment. — The remedies mentioned on p. 139 for tape worms are also useful for round worms. The remedy most commonly advised is to give 2 grains santonine for each bird. Dissolve this in water and use to mix the mash. As recommended on p. 141, all droppings should be collected and examined, also put out of reach of the birds.

Vale recommends the following: "Beat a new laid egg

with 1 tablespoonful of oil of turpentine and mix thoroughly by shaking. Give a teaspoonful of the mixture night and morning for a few days; or divide ¼ of an ounce of areca nut in powder, into 4 parts, and give 1 part each morning, fasting, with a dessertspoonful of sweet oil 2 hours after each powder."

Flukes

Flukes or trematode worms are small, flat and usually oval-shaped. Figure 24 gives a fair idea of the appearance of these parasites.

Regarding these parasites in poultry, Theobald [1] says:

"The Trematode worms or Flukes found in the fowl are 3 in number. One is found in the egg (*Distoma ovatum*) the others in the esophagus and intestines.

"The Fluke found in the esophagus of the fowl is known as *Cephalogonimus pellucidus*, as transparent reddish fluke about 9 mm.

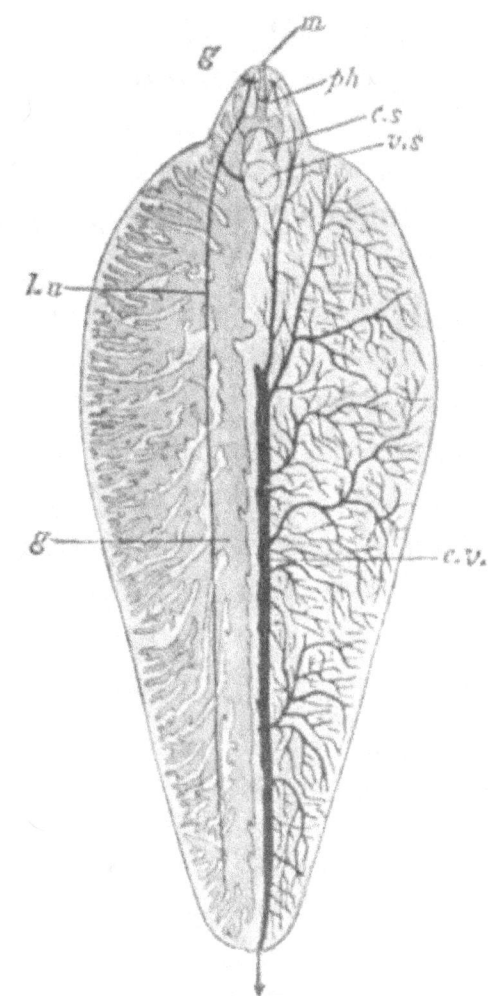

FIG. 24. — Trematode worm or fluke showing internal structure. (From Thompson after Sommer.)

long. These were found by Von Linstor and Railliet. In the intestines Neumann enumerates 7 species, namely, *Notocotyle triserialis*, *Distoma oxycephalum* Rud., *D. dilatatum* Miriam, *D. lineare* Zeder, *D. ovatum*, *D. armatum* Molin, and *Mesogonimus commutatus*

[1] Theobald, F. V., "Parasitic Diseases of Poultry." London.

Son. These, however, are not all distinct: *dilatatum* is undoubtedly the same as *oxycephalum*; *armatum* is also probably the same."

None of these trematode worms are of any pathological importance, although, as is well known, they often cause serious maladies in other animals. All the flukes that have two hosts undergo a complicated metamorphosis, the early stages always taking place in some water-mollusk. Those found in the fowl have not had their life-histories worked out.

Numerous flagellate and other microscopic parasites have been described from the intestine, ceca and cloaca of fowls. Martin and Robertson [1] mention particularly the flagellates, *Chilomastix gallinarum*, *Trichomonas gallinarum*, *Trypanosoma eberthi* and *Trichomastix gallinarum*.

Berké [2] describes a microfilaria occurring very abundantly in the liver of domestic fowls.

None of these parasites causes serious injury so far as known.

[1] Martin, C. H., and Robertson, Muriel, "Further Observations on the Cecal Parasites of Fowls with some Reference to the Rectal Fauna of Other Vertebrates." *Quart. Jour. Micros. Sci.* (London), N. S., Vol. 57, pp. 53–81, 1911.

[2] Berké, Centralb. f. Bakt., etc., Abt. 1, Bd. 58, pp. 326–330, 1911.

CHAPTER XI

Diseases of the Respiratory System

ANATOMY AND PHYSIOLOGY

The respiratory organs of birds are the nasal passages, the pharynx, larynx, trachea, lungs and air sacs. The form and general appearance of the lungs and trachea are shown in Fig. 25.

The respiratory apparatus differs somewhat in structure and function from that of mammals. As in mammals the trachea (windpipe) divides into the primary bronchi, one passing to each lung. In birds these bronchi do not divide and subdivide as in mammals, but each passes to the

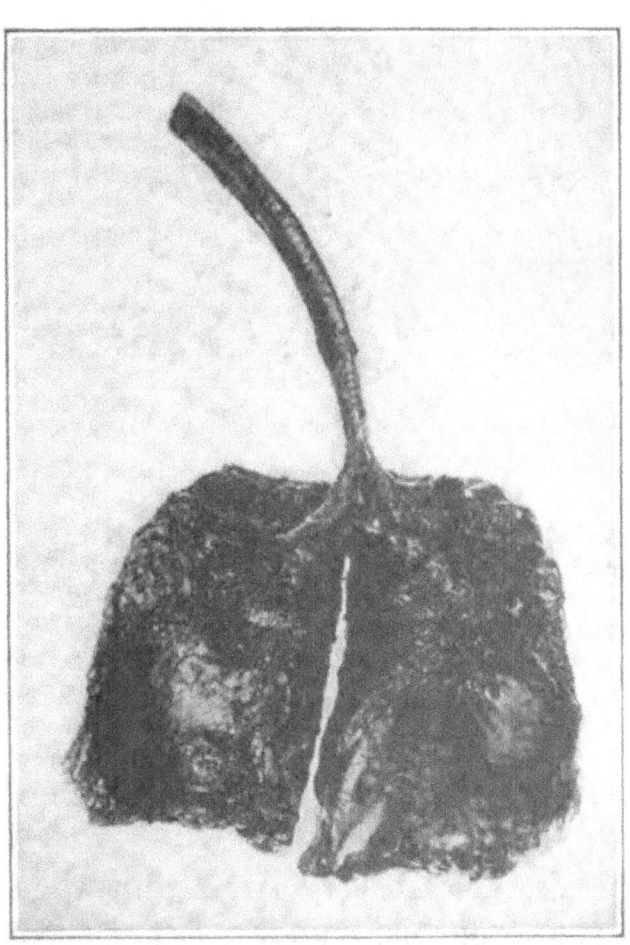

Fig. 25. — Photograph of the lungs of a domestic fowl. The air passages are injected with gelatine. (Original.)

posterior end of its lung where it opens into the abdominal air sac. This relation is shown in Fig. 26.

The primary bronchus gives off secondary bronchi which radiate toward the surface of the lungs. The secondary bronchi give off smaller radiating branches, the tertiary bronchi. Both primary and secondary bronchi remain of practically uniform diameter throughout their entire length. For the most part these tubes end blindly, but some of them communicate with the air sacs. This tubular system makes up the air-containing portion of the lungs. It is embedded in a network of almost naked blood vessels which make up the spongy tissue of the lungs. The aëration of the blood takes place through the walls of these vessels. The intimate relations of the tubular and vascular systems of the lungs are shown in Fig. 28.

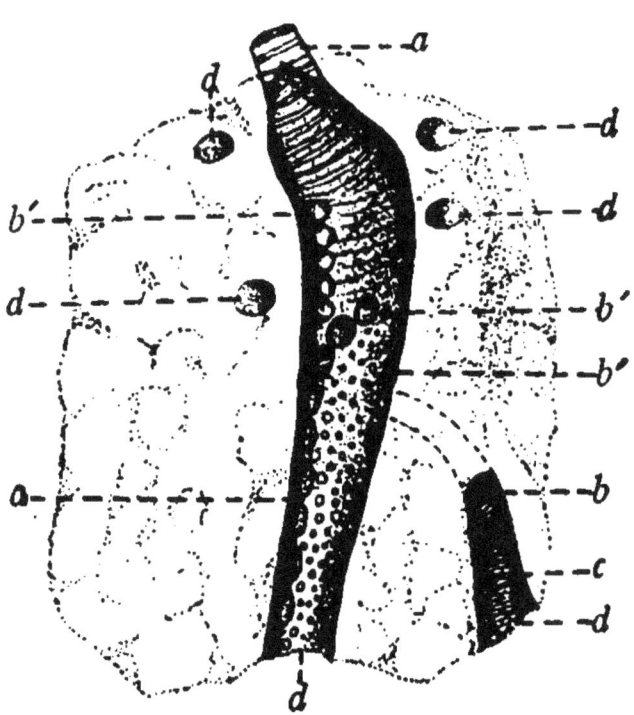

Fig. 26. — Diagrammatic drawing of the left lung of a domestic fowl. *a*, primary bronchus; *b*, secondary bronchus; *b²*, openings of other secondary bronchi; *c*, opening of tertiary bronchus; *d*, openings into air sacs. (Original.)

The air sacs are very large, thin-walled sacs which open into the bronchial tubes as described. When expanded with air these sacs fill all the available space in the thoracic-abdominal cavity and axilla. A small sac also lies along the ventral side of the neck, while diverticula from the large sacs are embedded among the muscles and even penetrate some of the bones. These sacs function chiefly as reservoirs of air. Some aëration of blood takes place in the sacs and they also help to reduce the relative weight of the body. The air sacs are the elastic or bellows-like portion of the respiratory apparatus. The lungs, on account of their structure and position in the body, are permanently distended. During inspiration the air passes through the trachea and lungs into the sacs. Fresh outside air is thus brought into the portion of the lungs where the blood is aërated. During expiration air from the sacs is forced back through the lungs. The current sucks out the air from the blind ending tubes and to some extent

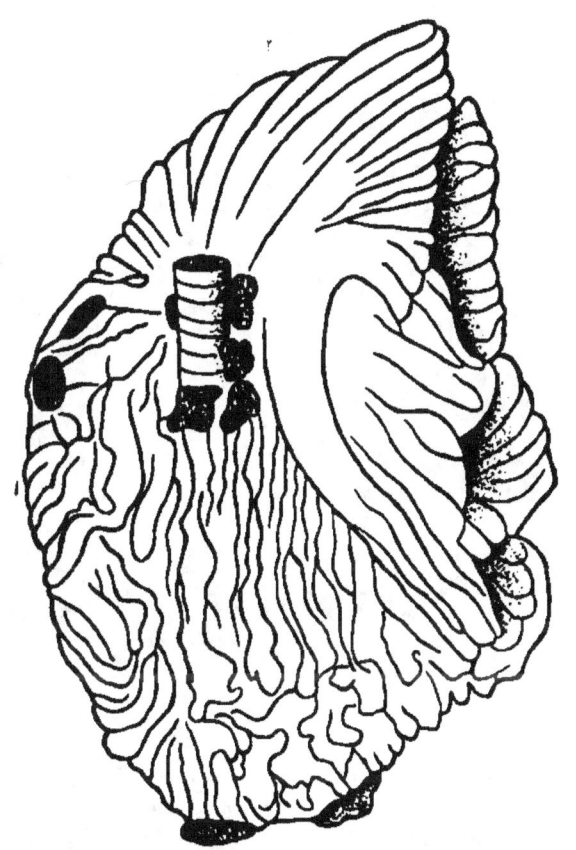

FIG. 27. — Ventral surface of the right lung of a fowl, injected with wax. (From Gadow, after Stieda.)

supplies comparatively fresh air in expiration. Thus the respiratory apparatus in birds is more efficient than in mammals, where fresh air is never available for the aëration of the blood, the entire process being carried on by residual air.

In addition to aëration of the blood, the respiratory apparatus eliminates most of the waste moisture of the body and is, therefore, the temperature regulator. In mammals this function is performed by the sweat glands and the secretion of the kidneys. Birds have no sweat glands and the secretion of the kidneys contains relatively little moisture.

The air passages are lined with mucous membrane and this membrane is the seat of several diseases. Diseases are easily

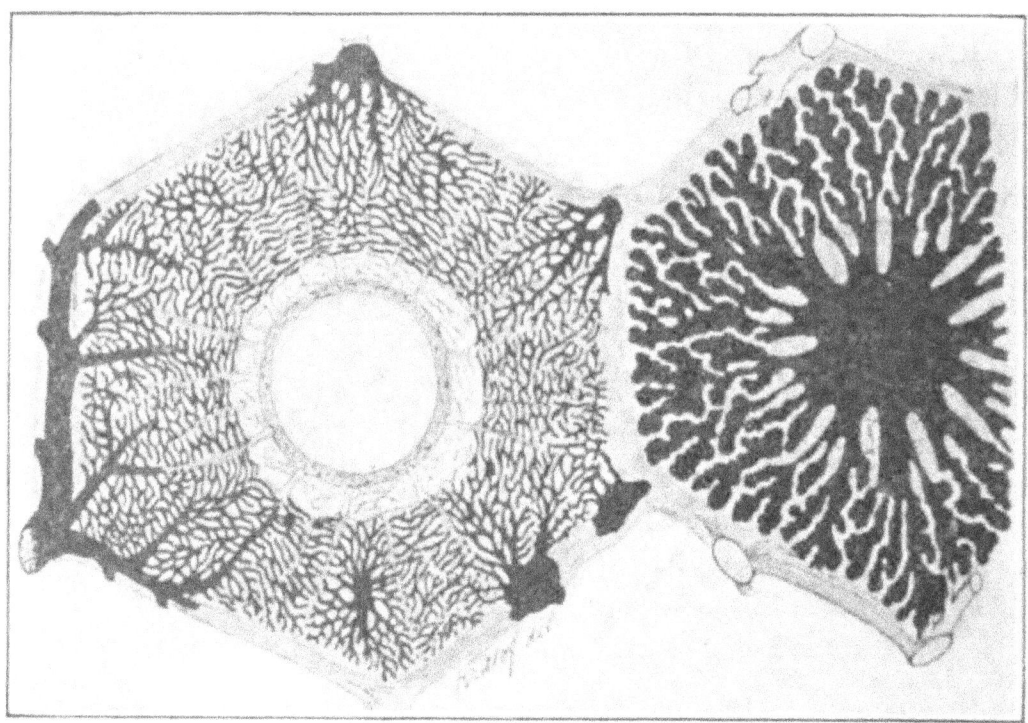

Fig. 28. — Cross section of two of the small air tubes of the lung of a goose. On the right the air passages are filled with black injection mass. On the left the arteries are similarly injected. (From Oppel, after Schulze.)

transferred from one part of the respiratory system to another, since the passages and also the lining membranes are continuous from the nostrils to the air sacs. There are also diseases of the vascular part of the lungs. Some of these diseases are caused by unfavorable conditions as exposure to cold, drafts of air, or moist air or to improper food. Others are due to specific organisms. Most of the latter are contagious. Exposure to unfavorable conditions also reduces the ability of the birds to resist infectious diseases.

Catarrh (Simple Catarrh; Non-contagious Catarrh; Cold)

One of the most common diseases of the air passages is catarrh (cold). It is often hard to distinguish this disease from early stages of roup and diphtheria. The characteristic symptoms of the latter disease should be carefully looked for, lest the flock become infected with a dangerous contagious disease. In cases where there is a suspicion of either of these diseases it is better to isolate the sick birds. Catarrh is non-contagious. It usually affects only a few individuals in the flock, but in cases of exposure of the flock to the unfavorable conditions which cause the disease it may occur in quite a number of birds at the same time.

Diagnosis. — Affected birds sneeze frequently, appear dull and lose their appetites. In early stages of the disease a thin mucous secretion is discharged from the nostrils. Often the eyes are watery and the eyelids inflamed and swollen. Later, in severe cases, the mucous secretions become gelatinous. The head passages may become entirely filled with this thick secretion. Breathing is then entirely through the mouth and is accompanied by a wheezing sound. A watery liquid drools from the mouth. Death occurs, apparently from exhaustion, in very severe cases.

Etiology. — The cause of catarrh is exposure to cold, to drafts of air, to damp atmosphere due to improper housing conditions, or to wet weather. Weak stock or improperly nourished birds are more likely to be affected by these conditions than strong, vigorous and well fed individuals.

Treatment. — With strong, healthy stock it is usually only necessary to remove the cause. Affected birds should be kept in warm, dry, well ventilated rooms. Daily individual treatment is effective and may profitably be applied if the attack is severe and the bird valuable.

This treatment when most effective involves three steps:

(1) Removal of secretions. The mouth and nostrils and the eyes, if affected, should be washed with warm water containing 1 teaspoonful of common salt to the quart. A small wad of absorbent cotton may be used to apply this cleansing solution. The sides of the head under the eyes and around the nostrils should be massaged gently to loosen the secretion.

(2) Disinfecting the air passages and eyes. The air passages and eyes should now be disinfected with one of the following solutions, given here in order of preference:

a. Potassium permanganate, 2 per cent solution.
b. Boracic acid, 3 per cent solution.
c. Creolin, 1 per cent solution.
d. Hydrogen dioxide and water, equal parts.
e. Carbolic acid, 2 per cent solution.

These solutions may be injected into the nostrils with a small syringe or a medicine dropper, but on account of the small aperture of the nostrils they are more effectively applied through the internal opening of the air passages. This is a long, widely open slit in the roof of the mouth. It is easily exposed in a position to receive the treatment by holding the bird head down, grasping the head, comb down, in the hand and opening the mouth with the thumb. The solution may then be injected into the slit or poured in with a teaspoon. The head should be held firmly in this position for several seconds after the treatment to allow the solution to penetrate to all parts of the head passages. Gently massaging the sides of the head also helps distribute the disinfectant. The eyes may be washed with the same disinfecting solution used for the nostrils.

(3) Applying oil to the head passages. A quarter of a teaspoonful of oil of thyme, oil of eucalyptus, or even sweet oil should now be administered in the same way as the disinfecting solution. If the eyes are affected introduce 2 drops of 15 per cent argyrol solution.

When the head passages are not filled with mucus the application of the oil to the nostrils and argyrol to the eyes without the previous steps is very beneficial.

Prognosis. — A great majority of the birds recover in a few days if the cause is immediately removed. If the cause continues to act they may become worse and die, or the disease may become chronic and persist for a long time.

Bronchitis, Croup

This disease may follow catarrh as a direct extension of the inflammatory processes in the membrane of the nasal cavities and throat to the mucous membrane of bronchial tubes.

Diagnosis. — The symptoms of bronchitis are the symptoms of a hard cold (severe catarrh) with rapid breathing and cough. It may be distinguished from a cold by the peculiar sounds made in breathing. In the early stages of the disease this is a whistling sound made by the passage of the air over the dry, thickened membrane. As the disease advances mucus collects in the tubes and the breathing is accompanied by a rattling or bubbling sound. Under favorable conditions the symptoms do not usually pass beyond this stage but soon disappear. In very severe cases the birds become very sleepy and refuse to eat. The wings droop. The feathers are roughened and breathing becomes more and more difficult, until finally the bird dies. The less severe forms of the disease may become chronic, while the symptoms of rattling breath and coughing up mucus may persist for a long time. In this form of the disease the birds appear well except for the above symptoms.

Etiology. — When it follows a hard cold, bronchitis may be caused by an extension of the inflammation of the mucosa of the throat to the mucosa of the bronchial tubes. It may also be caused directly by exposure to cold, drafts, and dampness; or it may result from irritation of the mucous membrane caused by inhaling irritating vapors, dust or foreign particles.

Treatment. — Place the patient in a warm, dry, well ventilated but not drafty room. Feed bread or middlings moistened with milk, and add to this food 2 grains of black antimony twice a day. A demulcent drink is often beneficial. A very good one is made by steeping a little flax seed in water. Other demulcent drinks are made by dissolving honey or gum arabic in water. This treatment is sufficient for mild cases. Salmon recommends the following treatment for severe attacks: "If the attack promises to be severe, it may sometimes be checked in the early stages by giving 10 drops of spirits of turpentine in a teaspoonful of castor oil and repeating this dose after 5 or 6 hours. It should not be continued after there are signs of purging, for fear of exhausting the strength of the patient. In the very acute cases, where the whistling or snoring sounds with the respiration indicate a croupous form of inflammation, and where the gasping shows great obstruction of the air passage, relief may be obtained by giving from 3 to 6 drops of either the sirup or the wine of ipecac.

"Medicines should be administered very carefully in diseases affecting the trachea and bronchi, as otherwise they may enter the air passages and increase the irritation."

Prognosis. — In the ordinary and chronic forms the birds usually recover. In the more severe forms a large per cent of the affected birds die.

Influenza (Epizoötic, Grippe, Distemper)

The symptoms are the same as those of a severe cold accompanied by fever and usually also by diarrhea. It appears to be a contagious germ disease as it often affects a large number of birds kept together.

Quite possibly this is not a separate disease but is either a severe form of cold occurring in many individuals of a flock which has been exposed to unfavorable, unsanitary conditions, or is a mild form of roup. Affected birds should be isolated. If diarrhea is present give one-half to one teaspoonful of Epsom salts dissolved in water. Treatment the same as for colds.

Prognosis. — Most cases recover in a week or ten days. Severe cases die in a day or two or sometimes within a few hours after the appearance of the first symptoms.

Roup

Veterinarians have distinguished two diseases belonging to this general class of troubles as follows: (*a*) roup or contagious catarrh when only catarrhal symptoms are present, and (*b*) diphtheria, diphtheritic roup and canker when diphtherial patches and false membranes are formed. The bacteriologists Harrison and Streit,[1] consider these different stages of the same disease. This view has been quite generally accepted. Cary[2] and several other workers[3] not only consider these as one disease but also believe that sorehead, chicken pox or *epithelioma contagiosum* is also a form of this disease. Evidence for the identity of "avian diphtheria" or "diphtheritic roup" and chicken pox is

---

[1] Harrison, F. C., and Streit, H., "Roup." Ontario Agr. Col. and Expt. Farm Bull. 125. 1902. *Ibid.*, Bul. 132. 1904.

[2] Cary, C. A., "Chicken Pox or Sore Head in Poultry." Alabama Col. Sta. Bul. 136. 1906.

[3] For example, Kingsley, A. F., "Epithelioma Contagiosum." *Amer. Vet. Rev.*, 30. 1907. Hadley, F. B., and Beach, B. A., "Controlling Chicken Pox, Sore Head or Contagious Epithelioma by Vaccination." *Proc. Amer. Vet. Med. Assoc.*, Vol. 50, pp. 704–712. 1913.

cited by several European investigators.[1] On the other hand Haring and Kofoid[2] and Sweet[3] and Fally[4] believe that there is convincing evidence that sore-head is distinct from the roup diseases. These questions cannot be settled with the present knowledge of the causes of these diseases. The probability is that there are at least three diseases although each of them may under certain conditions produce lesions similar to those of either of the other two. In fact the disagreement in the results of different investigators suggests that there may be several contagious diseases which produce nearly identical lesions. In the present work nasal roup, diphtheritic roup, and chicken pox will be discussed as separate diseases.

Nasal Roup or Contagious Catarrh

The disease called "roup" by poultrymen is a contagious catarrh. It attacks principally the membranes lining the eye, the sacs below the eye (infra-orbital sinuses), the nostrils, the larynx, and the trachea.

Diagnosis. — The general characteristics of the disease have been very well described by Salmon. The symptoms first seen are very similar to those of an ordinary cold, but

---

[1] Carnwath, T., *Arb. K. Gsndhsamt. Orig.* 27, pp. 388–402. 1907. Schmid, G., *Centbl. Bakt., etc., Orig.* 52, pp. 200–234. 1909. Ratz, I., *Allotorvosi Lapok*, Vol. 33, pp. 184–186. 1910. Sigwart, H., *Centbl. Bakt., etc., Orig.* 56, pp. 428–464. 1910. Uhlenhuth and Manteufel. *Arb. K. Gsndhsamt.*, Vol. 38, pp. 288–304. 1910. von Betegh, L., *Centbl. Bakt., etc., Orig.* 67, pp. 43–50. 1912. von Katz, S., *Monatsch. Prakt. Tierheilk.*, Vol. 25, pp. 41–46. 1913.

[2] Haring, C. M., and Kofoid, C. A., "Observations concerning the Pathology of Roup and Chicken Pox." *Amer. Vet. Rev.*, Vol. 40, pp. 717–728. 1912.

[3] Sweet, C. D., "A Study of Epithelioma Contagiosum of the Common Fowl." *Univ. of Col. Pubs. Zool.*, Vol. 11, pp. 29–51. 1913.

[4] Fally, V. *Ann. Med. Vet.*, Vol. 57, pp. 68–75. 1908.

there is more fever, dullness and prostration. Harrison and Streit say that although the head is often very hot the body temperature is normal or only very slightly higher than normal. The discharge from the nasal opening is at first thin and watery, but in two or three days becomes thick and obstructs the breathing. The inflammation, which begins in

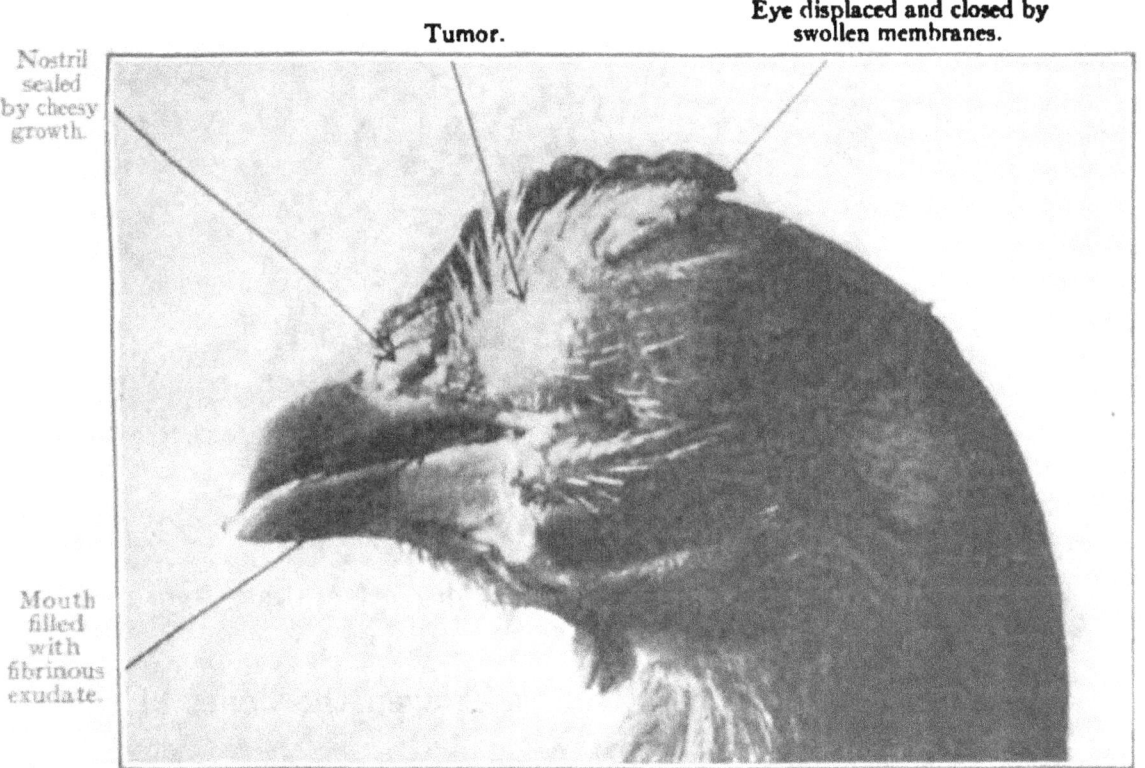

Fig. 29. — Photograph of a fowl's head showing infra-orbital tumor caused by roup. (After Roebuck.)

the nasal passages, soon extends to the eyes and to the spaces which exist immediately below the eyeballs. The eyelids are swollen, and are closed much of the time. They may be glued together by the accumulated secretion. The birds sneeze and shake their heads in their efforts to free the air passages from the thick mucus. The appetite is diminished and the birds sit with their heads drawn in, wings drooping, with the general appearance of depression and illness.

When the inflammation reaches the spaces or sacs beneath the eyes it causes the formation of a secretion very similar to that of the nose, and as this becomes thick it collects, distends the walls of these spaces, and produces a warm and painful swelling, which is seen just below the eyes and may reach the size of a hickory nut. This swelling presses with much force on the eyeball, which is displaced and more or

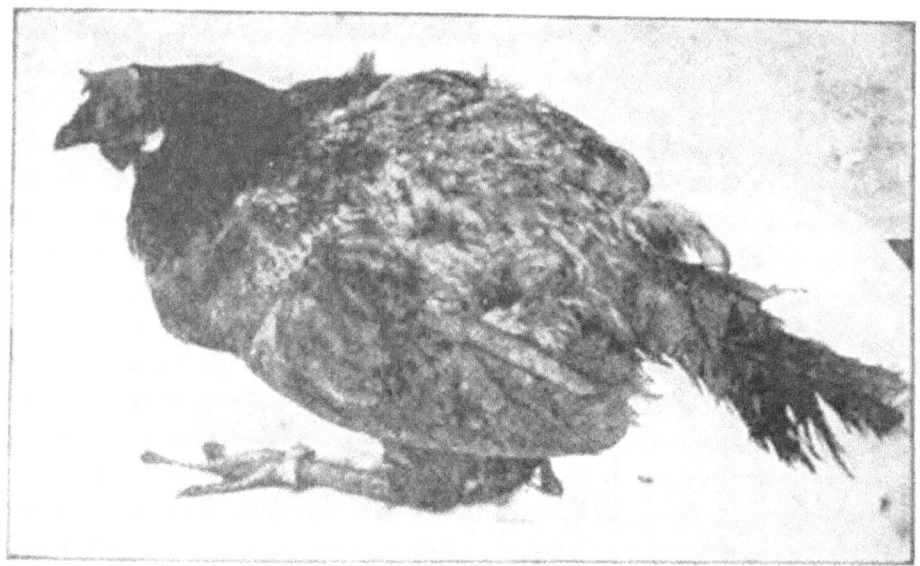

Fig. 30. — Showing appearance of a hen a day before death from roup. (After Harrison and Streit.)

less deformed; and in extreme cases even the bones of the head may give way before it.

The closure of the eyes prevents the badly affected birds from finding food; the accumulation of mucus in the nostrils completely obstructs these passages, so that the beak must be kept open in order to breathe; the obstruction of the windpipe and the smaller air tubes causes loud breathing sounds and difficult respiration.

In the severe and advanced cases the birds sit in a somnolent or semiconscious condition, unable to see or to eat; their strength is rapidly exhausted, and many of them die within a week or ten days. A part of the affected individuals recover, but others continue weak and have a chronic form of the disease for months; during which time they continue to disseminate the contagion.

This disease is distinguished from diphtheria by the absence of the thick, tough, and very adherent newly formed membranes (false membranes) in the nostrils, mouth, and throat which are characteristic of the latter.

The Course of the Disease. — The course of roup is usually of long duration. A simple, putrid discharge from the nose may stop in three or four weeks. But generally the symptoms last for months. When the eyelids become swollen and tumors appear, the case is usually chronic. Affected birds may be better for a few days or weeks, and then become very weak again. Damp, cold weather usually intensifies the disease.

It is well known that fowls may be more or less sick from roup for one or even several years and these birds should have the greatest care and attention, or else be killed at once, for they are generally the cause of new outbreaks. Once introduced, roup may remain in a flock for many years. The first cold and moist nights of the fall and early winter cause all kinds of catarrhs, which in many instances are followed by roup. Roup spreads rapidly in the winter time and may attack from 10 to 90 per cent of the fowls in a flock. Towards spring, the disease gradually disappears; during the summer months, a few birds remain chronically affected; and then the first cold nights give the disease a fresh start.

There is a great deal of difference in the susceptibility of fowls to the disease. Young fowls and finely bred fowls are especially liable to contract it. Some birds are apparently naturally immune and never take the disease. Others apparently have it in a mild form and completely recover, having thereby acquired an immunity.

Etiology. — Several organisms have been isolated from the lesions of birds suffering from roup.[1] Four of these have some claim to be considered the cause of the disease. These include three species of bacteria and one protozoön. There is also some evidence that the cause of the disease is an invisible virus. While the specific organism or organisms which cause the disease are not certainly known its infectious nature is well established. It is probably carried from one individual to another in a flock, by the particles of dried secretion in the air or possibly by the food and drink contaminated by the diseased birds. It may be introduced into a flock by the bringing in of birds from an infected flock, or by birds that have contracted the disease at shows. Possibly it is sometimes carried on the shoes or clothing of persons coming from infected yards or houses, and possibly also by wild birds or pigeons which fly from one poultry yard to another. While a source of infection is necessary for the production of the disease it does not appear to attack birds when the mucous membrane is in a healthy condition. It is most apt to attack birds that are suffering with catarrh. When a flock once becomes infected the birds which develop a mild chronic form of the disease serve as sources of infection whenever exposure to cold and dampness causes catarrh in the unaffected

---

[1] Much confusion in regard to etiology of the diseases belonging to general class commonly referred to as roup arises from the fact that several investigators believe them to be stages in a single disease. A partial list of the literature on the etiology of roup and diphtheria follows. Harrison, F. C., and Streit, H., *loc. cit.* Uhlenhuth and Manteufel, *loc. cit.* von Betegh, L., *loc. cit.* von Katz, S., *loc. cit.* Guerin, C., *Ann. Inst. Pasteur*, T. 15, pp. 941–952. 1901. Borrel. *Compt. Rend. Soc. Biol. (Paris)*, T. 57, pp. 642–643. 1904. Galli-Valerio. *Centbl. Bakt. Orig.*, Bd. 36, pp. 465–471. 1904. Streit, H., *Ztsch. Hyg. u. Infectionskrank.* 46, pp. 407–462. 1904. Muller, R., *Centbl. Bakt., etc., Orig.* 41, pp. 423–426; 515–523; 621–628. 1906. Bordet, J., et Fally, V., *Ann. Inst. Pasteur*, T. 24, pp. 563–568. 1910.

birds. Thus in infected flocks an outbreak of roup usually follows catarrh caused by exposure and this fact has led some poultrymen to think that the disease may be caused directly by exposure. In some flocks it appears annually with the cold damp weather of late autumn and breaks out again at every radical change of temperature and moisture conditions throughout the winter. Vigorous and properly nourished birds are better able to resist catarrh and consequently roup than those that are delicate and improperly fed.

Treatment. — The best treatment is prevention. The disease can be prevented by stopping all sources of infection. Some things to keep in mind are:

1. In introducing new birds always procure them from uninfected flocks.

2. Isolate all new birds and all birds that have been exhibited at shows for two or three weeks to make sure that they do not develop the disease.

3. Exclude from uninfected house and yards poultry and all other animals, including men, coming from those that are infected.

4. Do not use implements as hoes, shovels, etc., that have been used on infected premises.

5. Keep the birds in a good hygienic condition, well nourished and in dry well ventilated houses and roomy yards.

When the disease has been introduced into the flock careful precaution may prevent its spread.

1. Immediately separate from the flock any bird that shows symptoms of the disease.

2. Disinfect the yards and houses with a 5 per cent solution of carbolic acid or better the cresol solution described in chapter II. Remove the litter from the houses and disinfect freely. This disinfecting solution may also be followed by whitewash.

3. Use potassium permanganate in all drinking water.

4. Keep watch of the flock so that any new cases may be isolated at once.

5. *Burn or bury deep all birds that die.*

The disease is amenable to treatment but this treatment must be individual and requires a great deal of time. It must be continued once or twice a day for quite a long time. It is, therefore, very expensive and consequently impracticable for ordinary stock. Moreover birds apparently cured are likely to become the source of infection for later outbreaks.

In the case of valuable show birds treatment may perhaps be advisable.

The treatment recommended for catarrh or cold may be used effectively in early stages of roup.

Harrison and Streit [1] give the following methods of treating roup:

"The germs of roup are not very resistant; they can easily be destroyed when present in cultures, or somewhere outside the animal; but in the animal tissue, they are very difficult to kill, because they penetrate into the tissue; and unless this too is killed, the germs continue living for a long time."

"Roup may be cured by remedies, if the treatment is careful and judicious. . . . If the eyes and nose are attacked, they have to be carefully washed, at least twice a day, with an antiseptic solution, such as 2 per cent boracic acid in a decoction of chamomile flowers, or ½ per cent solution of corrosive sublimate. Thus the micro-organisms are killed or at least, the diseased products which are discharged are removed, and the irritation caused by them; also the transformation into large cheesy masses is prevented.

"We had chickens badly affected with roup of the eyes, which were cured with boracic acid and chamomile. On

---

[1] Harrison, F. C., and Streit, H., Roup. Ont. Agr. Coll. & Exp. Farm. Bul. 125, Dec. 1902, pp. 1–16.

account of the smallness of the nostrils and nasal canals, it is very difficult to get the antiseptic solutions into the nose and nasal cavities; but it can be done with a small syringe. If this treatment is too troublesome, then the nostrils, at least, should be washed and opened several times a day, to allow the secretions to pass away. We have treated chickens for 14 days by daily washing with a $2\frac{1}{2}$ per cent solution of creolin and glycerine. After the washings, small plugs of cotton wool, filled with mixture, were placed in the nostrils and lachrymal ducts. This remedy did not cure the roup, although the same mixture readily kills the roup bacillus in cultures in from 2 to 3 minutes. The greatest hindrance to a sure cure by remedies which have been used locally, is the ability of the germ to penetrate into the tissue and the many secondary cavities of the nostrils which cannot be reached by the antiseptic.

"Another method of treatment which gives excellent results, especially in the early stages of roup, is the use of 1 to 2 per cent of permanganate of potash. Fowls are treated in the following manner: The nostrils are pressed together between thumb and forefinger in the direction of the beak two or three times. Pressure should also be applied between nostrils and eyes in an upward direction. This massage helps to loosen the discharge in the nostrils and eyes. The bird's head is then plunged into the solution of permanganate of potash for 20 or 30 seconds, in fact the head may be kept under the solution as long as the bird can tolerate it. The solution is thus distributed through the nostrils and other canals and has an astringent and slight disinfecting action. This treatment should be given twice a day and continued until all symptoms have disappeared.

"If there are solid tumors in the eyelids, they should be opened so that the skin may bleed freely. The cheesy matter should be removed and the surrounding membrane touched with a 5 per cent carbolic acid or silver nitrate solution, and then a cotton plug put in again to prevent the cavity from healing too quickly. We have cured chickens in this way in about a fortnight.

"As all these methods of treatment demand a good deal of time and care, they cannot well be used for whole flocks, but the more valuable fowls may be treated in this manner. Farmers and poultrymen should first try the permanganate of potash method of treatment as it is the easiest to employ.

"Food remedies influence roup only by strengthening the fowls and assisting nature to throw off or conquer the disease."

The birds which are being treated should be kept in a dry, warm, well ventilated room with good nourishing food. The drinking water should be frequently changed.

Prognosis. — In infected flocks this disease caused a direct annual loss of 10 to 15 per cent of the flock. Also many birds contract a chronic form of the disease which affects them for months or years. Careful individual treatment will save the lives of many birds, but such treatment is economically inadvisable except in case of very valuable birds.

Diphtheritic Roup (Avian Diphtheria or Canker)

As previously stated this disease is considered by several investigators as a stage or a form of the same disease as nasal roup. There is, however, some good evidence[1] that they are

---

[1] For instance, in the Maine Station flock occasional cases of nasal roup appeared annually with unfavorable weather conditions, but there were *never* any false membranes formed. About five years ago some new stock was introduced. A few months later cases of typical avian diphtheria appeared in these birds and a little later in other birds in the same pens. This disease was sometimes associated with the lesions common to nasal roup, but often the two diseases were quite separate.

separate diseases although in early stages they cannot be distinguished. There has also been a considerable discussion of the possible identity of avian and human diphtheria, but the evidence that these are distinct seems conclusive. Diphtheritic roup is distinguished from nasal roup by the formation of false membranes on the mucous surface of the nostrils, eyes, mouth, throat, trachea or bronchi. These membranes are a tough, grayish or yellowish growth and adhere very firmly to the underlying tissue.

The first symptoms appear in from three to five days after exposure to contagion. The duration of the disease varies from a few days to several months. Many birds in the flock appear to be naturally immune or sufficiently vigorous to overcome the disease without the formation of the characteristic lesions. On account of mild, undetectable cases diphtheritic roup is very hard to eradicate without sacrificing the whole flock, disinfecting the premises, and starting anew with incubator chicks or clean purchased stock.

Diagnosis. — Following the excellent account given by Salmon it may be said that diphtheria begins as a local irritation or inflammation at some point on the internal surface of the mouth, throat, nostril or eyes. At this time the general health is not yet affected, and there is nothing but the diphtheritic deposit to indicate that the bird has been attacked. This deposit is at first thin, yellowish or whitish in color, and gradually becomes thicker, firmer, and more adherent, so that considerable force is required to remove it. The mucous membrane beneath the deposit is found, when the latter is removed, to be inflamed, ulcerated, and bleeding, but it is soon covered by a new deposit. This deposit is called a false membrane, and when it is situated where the air passes over it in breathing it dries, becomes uneven and fissured, and its color changes to a dark brown.

While the false membranes over the parts first affected are

becoming thicker, the inflammation extends to the adjoining surfaces, and new diphtheritic centers develop, uniting with each other until the cheeks, the tongue, the palate, the throat, and the inside of the nostrils are covered. Very often the

Fig. 31. — Diphtheritic roup or canker. (After Roebuck.)

inflammation extends from the nostrils to the eyes and the sacs beneath the eyes, and sometimes it penetrates the air tubes to the lungs or the gullet to the crop.

This extension of the disease leads to the appearance of other symptoms. The inflammation in the nostrils causes sneezing and the escape of a thin, watery secretion from the nasal openings; the thick false membranes fill up the nasal passages and the throat and obstruct the breathing; a thick, viscid secretion collects on the eyelids and glues them together; the sacs under the eyes fill up, and swellings are caused which disfigure the head; the poison which is produced by the growth of the microbe beneath the false membranes is absorbed and affects the nervous system, causing dullness, depression, and sleepiness. The affected bird stands with the neck extended and the beak open to facilitate the entrance of air into the lungs, and from the corners of the mouth there hang strings of thick, tenacious, grayish mucus. A characteristic disagreeable odor appears when the membranes begin to form, and as they increase in mass it becomes much stronger and by the time the birds are in the condition described above it is very objectionable.

At this time, which may be three to five days from the appearance of the first symptoms, the condition is very serious. Swallowing is difficult or impossible; the breathing is so obstructed that hardly sufficient air can be inhaled to support life; the head is swollen; the eyes are nearly or entirely closed; the feathers of the head, neck, and breast are foul with decomposing secretions from the nostrils and mouth; there is considerable fever; an exhausting diarrhea sets in; there is rapid loss of weight; the comb and wattles become pale and cold; the temperature of the body finally sinks below the normal; and death soon follows.

When false membranes form in the gullet, crop, and intestines, there is a rapid aggravation of the symptoms, an intense diarrhea, and the escape of blood with the droppings. This type of the disease is more frequent with water fowl than other birds. Some fowls in a flock are resistant, and after a few days of illness make a rapid recovery. Others remain dull, weak, and thin in flesh, and may have more or less catarrh and difficulty of breathing for a long time.

Etiology. — The cause of this disease like that of nasal roup is still a disputed question.[1] All of the organisms reported as causing the former disease have also been credited by one or another investigator with producing diphtheritic roup also. Whatever the nature of the causal organism the disease is certainly strictly contagious. It never appears except as a result of infection from a previous case.

Methods of Infection. — Birds may be infected by polluted food, drinking water or litter and probably also by dust containing particles of the dried secretions. The disease may be introduced into a flock by introducing infected stock. Occasionally the infection may be carried on implements or on the shoes of persons coming from infected yards. The first symptoms appear in from three to five days after exposure to contagion.

Treatment. — The best treatment is of course prevention. The same safe-guards suggested under nasal roup are also effective against diphtheria.

Two general lines of treatment for diphtheritic roup have been used with more or less success. (1) Local treatment of the diseased parts with disinfectants, and (2) vaccine and serum treatments. The cost of application of either of these makes them at present economically inadvisable for ordinary stock.

(1) *Disinfectant Method.* — In early stages of the disease the painting of the diphtheritic patches with tincture of iodine is sometimes sufficient. If the false membranes persistently reappear, as they are very likely to do, they may be burned away with 50 to 75 per cent hydrochloric acid or with silver nitrate. Great care should be taken not to touch un-

---

[1] Cf. references cited *supra*, p. 160.

affected parts of the mucous membrane as such wounds are likely to be infected and thus become the seat of fresh patches. After thick false membranes are formed it is necessary to remove them gently before applying the disinfectants. After the drastic disinfectants have been applied to the lesions the throat, mouth and head passages may be disinfected with one of the reagents recommended for catarrh or nasal roup. The potassium permanganate treatment outlined on page 163 is the easiest to apply.

Some workers have used a spray of oil of thyme, oil of eucalyptus or kerosene oil and have reported favorably. Reidenbach [1] tested a large number of antiseptics against fowl diphtheria, among which were 50 essential oils. He found that ajowan oil possesses the strongest antiseptic action. This oil is obtained from the fruit of *Ptychotis coptica*, an annual plant which resembles caraway and has for its habitat Egypt, Persia and the East Indies. This oil is on the market in this country but is not ordinarily carried by druggists.

All of these local treatment methods are unsatisfactory in most cases because the germs causing the disease are embedded deep down in the tissue underlying the false membranes. It is therefore very difficult thoroughly to disinfect the lesions. The tissues injured by strong disinfectants seem especially adapted for the propagation of the surviving pathogenic organism.

(2) *Vaccine and Serum Treatments.* — An attack of avian diphtheria confers an immunity, the duration of which depends in part at least upon the severity of the attack.[2] The fact of this acquired immunity suggests the possibility of establishing an immunity with vaccine and of treating the disease with vaccine or serum. Within the last fifteen years

---

[1] Reidenbach, J., *Geflügel Ztg.* (Leipzig), Bd. 26, p. 116, 1910.
[2] Sigwart, H., *loc. cit.*

a number of experiments in this line have been conducted both in this country and Europe. Some workers have tried as a curative agent the antitoxin prepared for human diphtheria. Others have used serum prepared from small mammals and fowls which have recovered from the disease. Still others have attempted to establish an immunity in healthy fowls and also to cure diseased birds with vaccines prepared from cultures of the organisms isolated from the lesions. Different investigators disagree as to the effectiveness of each of these methods of treatment. The disagreement of their results may be due to the variation in virulence of the cultures with which they worked.

Hopeful results have been lately obtained in this country at Ohio State University and at Purdue. The following brief account of the preparation of the vaccine and the method of treatment used successfully at Purdue and on nearby farms is given by Philips.[1]

"Cures and methods of cure for roup are so varied and uniformly unsatisfactory that it was thought advisable to experiment with roup vaccine. This vaccine was first made at Ohio State University and proved reasonably satisfactory to them.

"The method of making vaccine is very simple. The first process is to take cultures from under the ulcers and grow them from 24 to 36 hours on neutral agar at a temperature of 37.5 degrees C. Then wash off the organisms in a sterile normal salt solution and attenuate them for an hour and a half in a water bath at 64 degrees C. If the vaccine is to be left standing it is advisable to make it one-half per cent acid with carbolic acid, as this acts as a preservative.

"The method of standardization is the most difficult part

---

[1] Philips, A. G., "A Preliminary Investigation with Roup." *Jour. of the Amer. Assoc. of Inst. and Invest. in Poultry Husbandry*, Vol. 1, No. 4, pp. 28–31, 1915.

and requires accuracy and careful manipulation, but after one becomes accustomed to it he can standardize by the turbidity of the vaccine with enough accuracy to be correct as is needed to be used on chickens. The turbidity may be compared to that of two drops of milk in ten cubic centimeters of distilled water.

"The average dose is one cubic centimeter given subcutaneously with a hypodermic syringe. The most convenient place to inject the vaccine is under the skin over the region of the breast. This need not be repeated in immunizing healthy birds, but in treating sick birds a second or third injection may be necessary every five days.".

This method of treatment is not yet on a secure scientific basis and it cannot be used in practice by the poultryman or farmer until all doubt of its efficiency is removed and a reliable vaccine prepared and put on the market. If this is ever accomplished, treatment with vaccine will be much cheaper to administer and much more efficient than the local disinfection of lesions.

Prognosis. — In very acute cases death may occur in two or three days. More often even in fatal cases the disease runs for two or three weeks. Recovery may be complete in two or three weeks or an individual may develop a chronic form of the disease which continues for several months. If untreated about half of the birds which contract the disease die.

Pip (Inflammation of the Mouth)

The term "pip" as used by poultrymen evidently does not represent a separate disease but is the result of mouth breathing due to closure of the nostrils by cold or catarrh. The mucous membrane of the mouth and tongue become hard and dry. This is especially true at the end of the tongue. Mucous discharge from the mouth often collects and dries on to this hardened skin at the tip of the tongue. Thus scab may crack partly away from the tongue exposing a raw surface.

Etiology. — It would appear to be the case that the symptoms above described originated from different causes in different cases. The trouble may be due to specific infection, though a particular organism has not yet been definitely isolated as the cause. In some cases the symptom is apparently purely physiological, arising from a failure of the mucus-secreting glands to function properly, owing to a lowered physiological condition.

Treatment. — The essential points in the treatment of this diseased condition is first to treat the primary cause (cold, catarrh, etc.). In removing the scale or "pip" gentle measures are to be followed, otherwise a raw surface likely to ulcerate, will be left. It is better to keep the scab wet with an equal mixture of glycerine and water. If the scab comes off leaving a raw surface this should be treated with disinfecting solutions (see p. 168) and then with glycerine.

Prognosis. — Pip is associated with diseases of the respiratory organs which are often serious and sometimes fatal. It is this association which is responsible for the general belief that it is a dangerous malady. In itself it is not serious unless the mucous membrane is torn away with the scab and the wound becomes infected.

Canker

Membranes formed in diphtheritic roup are sometimes called canker, but there are frequently found cheesy patches on the mucous membrane of the mouth or tongue which are not associated with roup. These growths are frequently, at least, the result of a traumatic injury to the membrane. Male birds frequently have canker where they have been picked in the mouth by other males when fighting. The growths are made up almost entirely of pus germs. These growths should probably be considered as suppurating wounds. An unhealthy condition of the mucous membrane of the mouth due to digestive disorders is sometimes accompanied by spots of canker.

A good treatment for canker is undiluted creolin applied with a cotton swab. The swab should be held against each sore for a short time. The whole surface of each patch should be treated. Another good treatment is to wash the sores with hydrogen peroxide 1 part and water 1 part.

Thrush

This term is also sometimes incorrectly applied to the false membranes of diphtheria, but there are at least two cases of true thrush on record. That is, in two cases microscopic examination has shown that the patches, which in both these cases were in the lower part of the esophagus and crop, were made up of spores and filaments of the fungus *Saccharomyces albicans*. This fungus causes thrush in children and calves. This disease may also occasionally affect the mouth. It is impossible to distinguish it from other diseases causing similar formations except by microscopic examination. The treatment is the same as for canker.

Aspergillosis (Mycosis of the Air Passages)

This is a very common disease of poultry, often mistaken for tuberculosis. In adult fowls it is a frequent cause of the condition known as "going light," while in young chickens it probably ranks next to white diarrhea as a lethal agent.

The discussion of this disease here relates primarily to adult fowls. Aspergillosis in young chicks is treated in Chapter XIX.

Diagnosis.—In early stages of the disease the bird appears normal. Later there is a loss of appetite and an abnormal thirst. The bird becomes inactive, standing with head down, eyes closed, wings dropped and plumage roughened. There is an increase in the rate of respiration and a rise in body temperature. The breathing becomes labored and is accompanied by a rattling sound caused by the vibration of the mucus which collects in the trachea and bronchi. Diarrhea sets in. The bird becomes emaciated. Death from toxæmia, exhaustion or sometimes from asphyxiation may occur in from 1 to 8 weeks.

Two types of lesions are found at autopsy. The first type is whitish or greenish yellow membranous patches on the mucous lining of the air passages. They are most often found in the trachea, bronchi, small passages of the lungs and large air sacs. Occasionally they also occur on the walls of the air sacs in the interior of the bones. Lameness with swollen and inflamed joints results from this condition. They are also sometimes found upon the mucous membranes of the alimentary canal. The mucous membrane underneath these patches is thickened and inflamed. The patch itself is a thick, fibrous, membrane-like mass which contains fungal filaments. These filaments bear spores at the surface of the false membranes. In fact these patches represent the free growth of the mold on the surface of the mucous membrane, having very much the appearance which it presents when growing outside of the body on dead organic matter. The greenish color of the diseased area is due to the greenish filaments of the mold or fungus growing upon its surface. The filaments are not all on the surface, however, but they penetrate deeply into the tissues, causing inflammation and swelling, which obstructs the respiration, and at the same time they apparently produce a poison, which causes the general depression and fever.

Lesions of the second type are whitish or yellowish tubercles resembling the tubercles of tuberculosis. They vary in size from the head of a pin to a large pea. The tubercles are embedded in the tissues of the walls of the air passages, in the lungs and sometimes also in the liver, spleen and kidneys. Each tubercle contains a growth of mold at the center which is inclosed by a wall of animal cells.

A certain diagnosis of aspergillosis requires the identification of fungus filaments and spores within the lesions. This is of course impossible during the life of the bird.

Etiology. — The disease is caused by molds of the genus *Aspergillus* which grow on the mucous membrane of the air passage. The four parasitic species in order of their importance are *Aspergillus fumigatus, Aspergillus nigrescens, Aspergillus glaucus, Aspergillus candidus*. The appearance of one of these molds, when greatly magnified, is shown in Fig. 32.

These molds and their spores occur on dead organic material like straw, grain, etc. They are inhaled in breathing or swallowed with the food. This being the case the importance of avoiding musty litter, and moldy or musty grain of all kinds is apparent. As with most other diseases the resistance of the individual against infection is here an important matter. Some fowls will be able to stand musty litter and grain without any harm, while others will promptly develop aspergillosis. When once present in a flock aspergillosis is probably transmitted from generation to generation through the eggs.

Treatment. — The disease is prevented by having clean, dry, well ventilated houses and avoiding the use of moldy litter or grain. Vigorous birds under sanitary conditions are fairly resistant.

Since the disease probably sometimes spreads from bird to bird, affected individuals should be killed or isolated and the bodies of dead birds burned. Salmon says that "sometimes affected birds may be saved by applying flowers of sulphur or tincture of iodine to the diseased patches seen in the mouth

Fig. 32. — *Aspergillus fumigatus*. Greatly enlarged. (After Mohler and Buckley.)

and throat, and causing the birds to inhale the vapor of tar water or turpentine. Tar water is obtained by stirring 2 tablespoonfuls of wood tar in a quart of warm water and letting the mixture stand for a few hours. Then the birds are taken into a closed room, where the tar water is poured, a small quantity at a time, on a hot brick or stone until the atmosphere of the room is well charged with the vapor."

This treatment is of value only when the lesions are of the open type first described and when these are located in parts of the air passages accessible to the fungicides.

Prognosis. — The disease in adult fowls is ordinarily not recognized as such until an affected bird comes to autopsy, at which time the prognosis is certainly extremely unfavorable. So far as concerns ridding a poultry plant of the disease, however, the outlook is favorable if energetic sanitary measures along the lines indicated above are applied.

Congestion of the Lungs

Congestion of the lungs is a distention of the blood vessels which make up the vascular portion of those organs. The pressure of these distended vessels may close the smaller air passages, or a vessel may burst, filling the bronchi. In either case the patient soon suffocates.

Diagnosis. — The symptoms of this disease are difficult rapid breathing, sleepiness and an indisposition to move. A bloody mucus sometimes flows from the mouth. The comb is dark red or bluish from lack of oxygen in the blood. Symptoms appear suddenly and death occurs within a few hours.

Etiology. — This disease is caused by chilling the surface of the body. This contracts the surface vessels and a large volume of blood is sent to the internal organs. The pressure on the small elastic vessels of the lungs is too great and they either close the air passages by pressing against them or the vessel walls are ruptured by the internal pressure and the air passages become filled with blood. This disease most often occurs in denuded birds (hens during molting or young birds which have failed to feather out) or small chicks which have been exposed to cold or allowed to run out in cold, wet weather.

Treatment. — The rapid course of the disease makes treatment impracticable. Prevention is the only cure. Birds should be well nourished with plenty of green food and should be especially protected from cold and wet when molting. Also chicks which are in a stage between down (chick) and juvenal feathers need special protection. This disease often attacks brooder chicks and indicates something wrong with the brooding. The cause should be immediately sought out and removed, or considerable loss will follow from continued exposure of the flock.

Prognosis.— This disease is usually fatal in a few minutes or hours after its symptoms are noted. According to Salmon the patient sometimes develops pneumonia.

Pneumonia

This disease is a step beyond congestion of the lungs. The vessel are not only distended but liquid escapes through their walls and coagulates in the air spaces. The lung of a chick dead of pneumonia is dark colored and firm and heavier than water. A normal lung floats but a lung filled with this coagulated serum sinks.

Diagnosis. — The symptoms resemble those of congestion of the lungs. The bird stands with its head drawn back, its wings down, and its plumage ruffled. The comb is usually very dark. Respiration is rapid, labored and apparently painful. There is a loss of appetite, and abnormal thirst and constipation. A thick, adhesive, and often bloody mucus is sometimes discharged from the nostrils and mouth. In such cases there is usually coughing.

Post-mortem examination shows that either the bronchi are nearly filled with thick mucus or the smaller air passages of the lungs are filled with clotted serum. The first form of the disease is called broncho-pneumonia. In these cases death from suffocation has taken place while the vascular area of the lungs is little affected. This is the form of the disease which usually follows bronchitis. There is usually also some clotted serum in the smaller passages. The second form is known as croupous pneumonia. This form usually follows congestion of the lungs and often develops very soon after severe chilling. The lungs are dark in color and solid. The blood vessels are filled with blood and the air spaces are filled with clotted serum. A piece of such a lung will sink when dropped into water.

Etiology. — The cause of pneumonia in birds is not known. The similarity of this disease to human pneumonia, both in the circumstances of its occurrence and in its lesions, has led to the inference that the cause is also similar. The disease is probably caused by a bacterium which is very often present in the normal air passages but which is harmless except when there is congestion.

Treatment. — Ordinarily it will not be advisable to treat this disease. *A cure is unusual and involves such an amount of care and nursing as to make it a most unprofitable proposition.* The disease can be guarded against by keeping the flock in good condition and preventing exposure. Some cases will recover if removed to a warm well ventilated room and fed milk or raw eggs supplemented later, if the bird improves, with cod liver oil and chopped beef.

Various medicines, especially stimulants, have been recommended. One drop of tincture of aconite every two hours given in egg or milk is said to relieve the hard breathing. As a stimulant 2 drops of spirits of camphor and 10 drops of brandy in a teaspoonful of warm milk may be given 3 or 4 times a day. If the comb becomes dark and the bird is evidently failing rapidly add one drop of tincture of digitalis to the above medicine.

To reduce the congested condition of the lungs a counter irritation may be applied to the skin of the back over the region of the lungs. This may be done by raising up the feathers and painting the skin with tincture of iodine.

Prognosis. — This disease is more easily prevented than cured. A number of cases recover if placed in a warm dry room. Treatment is of little value. A large percentage of the birds die.

Tuberculosis

This disease is discussed in Chapter IX. It is primarily a disease of the abdominal viscera. Lesions occur also in the lungs in about one case in five. The infection of the lungs is usually of secondary origin.

The Air-Sac Mite

A species of mite (*Cytodites nudus*) infects the air-sacs and bronchi of poultry.

Diagnosis. — When the birds are not badly infested there are no external symptoms. If badly infested the bird may become anæmic and listless and finally die of exhaustion. Or, if the air passages are seriously obstructed by the collection of parasites and mucus, there will be a rattling in the throat and coughing, and death may result from suffocation. The presence of the parasites is often found only on examination of dead birds. They appear as a yellow or white dust, each particle of which is a mite. If closely watched the particles may be seen to move.

Etiology. — The mite probably enters the air-sacs by crawling in the nostrils and finding its way down the trachea and bronchi to the sacs. The parasites are able to live only a short time outside the bird's body. The mouth parts of these mites are modified into sucking tubes. They attack the mucous membrane of the air-sacs and bronchi. When the number of parasites is small they cause no serious inconvenience to the bird. When there are a large number present they may cause inflammation of the membrane and secretion of mucus or they may seriously obstruct the air passages.

Treatment. — Treatment of infested birds is probably useless. Feeding sulphur with the food or compelling the birds to inhale the fumes of burning sulphur or burning tar or the steam of boiling tar water has been recommended.

If a flock is infested with this parasite it is best when possible to start a new flock with incubator chicks raised on a new range and carefully protected from infection from the old flock. Or stock may be purchased from an uninfested flock. Birds with this parasite should not be bought or sold for breeding or laying purposes, as the flock into which they are introduced will become infested

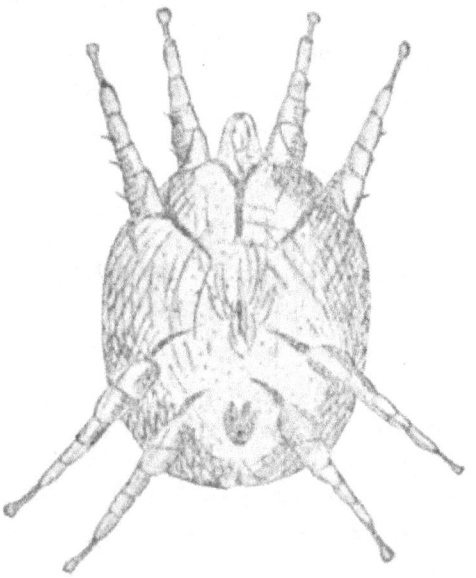

Fig. 33. — *Cytodites nudus.* The air-sac mite. Greatly enlarged. (After Theobald.)

from them. Since this parasite cannot live long outside the bird's body, the houses, runs, etc., do not remain infested long after all the diseased birds have been removed.

Prognosis. — A bird once infected is probably never free from the parasite but may live a long time little harmed by its presence.

CHAPTER XII

DISEASES OF THE CIRCULATORY SYSTEM

THERE are two classes of diseases treated in this chapter: (1) diseases of the organs of circulation (the heart and blood vessels), and (2) diseases of the blood.

DISEASES OF THE ORGANS OF CIRCULATION

Pericarditis (Inflammation of the Pericardium, Dropsy of the Heart Sac)

This disease is often found associated with other diseases of the circulatory system and with diseases of the lungs and air-sacs and also with soreness of the joints.

Diagnosis. — A differential diagnosis of this disease during life is not usually possible. Salmon gives the following symptoms: "There is great weakness, difficult breathing, the head being thrown backwards, and the breath drawn through the mouth in order to obtain sufficient air. If forced to run the bird soon falls. In a case observed by Hill there was tumultuous action of the heart and occasional spasms." Examination of a bird dead from this disease shows the heart sac full of serous liquid and sometimes the cavity is divided by false membranes which may attach to the heart as well as to the pericardium.

Etiology. — The causes of this inflammation are not known. It may result from exposure to cold or dampness.

Treatment. — Treatment is impossible since the disease

cannot be diagnosed until after death. Successive cases in the same flock indicate exposure of the flock to cold or wet weather or to confining the birds in insanitary houses. These conditions should be remedied. Salmon also recommends in such cases "2 to 4 grains of bicarbonate of soda to each bird daily in the drinking water."

Endocarditis (Inflammation of the Internal Membranes of the Heart)

In the examination of dead birds it is sometimes found that the membrane lining the heart is reddened and coagulated lymph may adhere to it. Little is known of this disease in fowls. It cannot be distinguished from pericarditis except by an examination of the heart. The cause and treatment suggested for that disease probably apply equally in these cases.

Myocarditis diphtheritica

According to Zürn, Bollinger has described a bacterial disease of the heart and blood vessels of fowls and pigeons. The disease is caused by a bacterium which resembles the bacterium of roup. The disease attacks the lining membrane of the heart and blood vessels, causing inflammation and the breaking down of the tissue. It especially affects the valves of the heart and aorta, where round or oval colonies of the bacteria are found on the membrane. In these patches fibrin and red and white corpuscles are mingled with the organisms. The walls of the small vessels of the lungs, liver, spleen, kidneys and intestines are also affected. The liver, spleen, and kidneys are enlarged. The bacteria are numerous in these organs as well as in the blood.

Little is known of the frequence of the occurrence of this disease and nothing of methods of treatment.

Enlargement of the Heart (Hypertrophy)

The heart of a fowl is sometimes enlarged. According to Cadeac this enlargement most frequently affects the right side of the heart. The muscle may be fatty and degenerate.

Diagnosis. — The distinctive symptom of this disease is a very rapid beating of the heart.

Etiology. — The cause of this hypertrophy of the heart muscle is not known, but it is probably due to some derangement in the nutrition of the muscle. The palpitations are increased by excitement or fright.

Treatment. — The disease is not usually recognized while the bird is alive. Treatment is therefore not possible.

Prognosis. — A hypertrophied heart may function for a long time. The violent beating may cause rupture of a blood vessel; sometimes several vessels are ruptured at the same time.

Rupture of the Heart and Large Blood Vessels

Internal hemorrhage due to the rupture of the heart or large blood vessels often occurs in full blooded fowls.

Diagnosis. — The bird becomes weak and drowsy, passes into a comatose condition and dies with the characteristic appearance associated with bleeding to death.

Etiology. — In full blooded fowls any excitement or overexertion which causes an increase in the rate of heart beat and an increased blood pressure may result in a rupture of the heart or one of the large vessels.

Treatment. — The accident cannot be predicted and treatment is impossible.

Prognosis. — The bird dies in a short time.

Thrombosis

This disease is characterized by the clotting of the blood in the great blood vessels and sometimes also in the heart. Sometimes the corpuscles settle out of the serum so that a part of the clot is clear.

Diagnosis. — This condition is not capable of diagnosis except at autopsy. Birds which show this condition, however, are often those which have been sick several weeks. They are usually in poor flesh and a gradual loss of appetite is often noted for some weeks before death occurs.

Etiology. — The cause of this disease is unknown.

Treatment. — As the disease is only recognized at autopsy no treatment is possible.

Leukæmia

Various cases of an alteration in the number of white corpuscles in the blood of fowls have been described. According to Warthin[1] in normal hen's blood the proportion of red blood corpuscles to white is 105–225:1, and only 14 per cent of the white cells are large lymphocytes, while in leukæmia of fowls the proportion of red to white cells may be less than 2 to 1 and a differential count of the white cells shows that there may be 84.5 per cent large lymphocytes. The tissue changes consist in tumorous nodules and infiltration of lymphoid cells in the liver, spleen, bone marrow and other organs.

From the literature it appears that investigators have found several different blood diseases which show the blood picture described above. Hirschfeld and Jacoby[2] and

---

[1] Warthin, A. S., "Leukemia of the Common Fowl." *Jour. Infect. Diseases*, Vol. 4, No. 3, pp. 369–381, 1907.

[2] Hirschfeld, H., and Jacoby, M., *Berlin. Klin. Wchnschr.* Bd. 46, pp. 159, 160, 1909.

Burckhardt [1] have found such a condition associated with the presence of tubercle bacilli in the blood. This was found both in spontaneous and experimental cases of tuberculosis.

Ellerman and Bang [2] found this condition in cases which were transmissible to other fowls by a filterable virus injected intravenously. This disease ran a chronic course. The typical leukæmic condition was reached about three months after the inoculation.

The only disease of this group which seems to be of any economic importance is infectious leukæmia or fowl typhoid, first described by Moore.[3] The relation of this disease to human typhoid is discussed on page 111.

Infectious Leukæmia or Fowl Typhoid

This is a bacterial disease often mistaken for fowl cholera but caused by a different species of bacteria and the lesions produced are somewhat different.

Diagnosis. — The following symptomatology is quoted from Moore: "From the statement of the owners of the fowls in the different outbreaks and from the appearance of those in which the disease was artificially produced, little can be positively recorded concerning the distinctive or characteristic symptoms. The only fowl examined antemortem from the natural outbreaks was first seen only a few hours before death, when it was unable to stand. If held in an upright position, the head hung down. There was a marked anæmic condition of the mucosa of the head.

---

[1] Burckhardt, J. L., *Ztschr. Immunitätsf. u. Expt. Ther.* Bd. 14, pp. 544–604, 1912.

[2] Ellerman, V., and Bang, O., *Ztschr. Hyg. u. Infektionskrank.* Bd. 63, pp. 231–272, 1909.

[3] Moore, V. A., "Infectious Leukœmia in Fowls — a Bacterial Disease Frequently Mistaken for Cholera." Ann. Rept. Bur. An. Ind., 1895–1896, pp. 185–205.

It had an elevation of nearly 3 degrees of temperature. An examination of the blood showed a marked diminution in the number of red corpuscles and an increase in the number of white ones. In the disease produced artificially by feeding cultures of the specific organism there was in most cases a marked drowsiness and general debility manifested from 1 to 4 days before death occurred. The period during which the prostration was complete varied from a few hours to two days. The mucous membranes and skin about the head became pale. There was an elevation of from 1 to 4 degrees of temperature. The fever was of a continuous type.

"Although the course of the disease in the different fowls was usually constant, there were many variations. In a few individuals the time required for fatal results was from 2 to 3 weeks, but ordinarily death occurred in about 8 days after feeding the virus, the rise in temperature being detected about the third day and external symptoms about the fifth or sixth, occasionally not until a few hours before death. The symptoms observed in the cases produced by feeding correspond with those described by the owners of affected flocks."

Moore found the only constant lesions to be in the blood and liver. The change in the blood as noted above was a decrease in the number of red and an increase in the number of white cells. The change in the liver is described by Moore as follows:

"The liver was somewhat enlarged and dark colored, excepting in a few cases in which the disease was produced by intravenous injections. A close inspection showed the surface to be sprinkled with minute grayish areas. The miscroscopic examination showed the blood spaces to be distended. The hepatic cells were frequently changed, so that they stained very feebly, and not infrequently the cells were observed in which the liver cells appeared to be dead and the intervening spaces infiltrated with round cells. The changes in the hepatic tissue are presumably secondary to the engorgement of the organ with blood."

Dawson's diagnosis of the disease (An. Rep. Bur. An. Ind., 1898, p. 350) differs somewhat from the one given by Moore.

It is very difficult to distinguish this disease from fowl cholera except by identifying the bacteria which produce the diseases. Moore contrasts the characteristic lesions in the appended columns:

Fowl cholera	*Infectious leukæmia*
1. Duration of the disease from a few hours to several days.	1. Duration of the disease from a few hours to several days.
2. Elevation of temperature.	2. Elevation of temperature.
3. Diarrhea.	3. Diarrhea very rare.
4. Intestines deeply reddened.	4. Intestines pale.
5. Intestinal contents liquid, muco-purulent, or blood stained.	5. Intestinal contents normal in consistency.
6. Heart dotted with ecchymoses.	6. Heart usually pale and dotted with grayish points, due to cell infiltration.
7. Lungs affected, hyperæmic or pneumonic.	7. Lungs normal, excepting in modified cases.
8. Specific organisms appear in large numbers in the blood and organs.	8. Specific organisms comparatively few in the blood and organs.
9. Blood pale (cause not determined).	9. Blood pale, marked diminution in the number of red corpuscles.
10. Condition of leucocytes not determined.	10. Increase in the number of leucocytes.

Attention should be called to the fact that as yet there seems not to have been a careful study of the condition of the blood in fowl cholera. Salmon observed many changes in this fluid which may have been similar to or identical with those herein recorded.

On page 201 of Moore's paper he gives the method of differentiating the two bacteria. This is, of course, dependent on microscopic examination and cultural tests. A full description of *Bacterium sanguinarium* is given by Moore on pages 188-191 of the paper cited above.

Etiology. — The disease is caused by a non-motile, rod-shaped bacterium (*Bacterium sanguinarium*). This bacterium causes the disease when injected into the blood or when fed. In a few cases fowls are known to have contracted the disease by picking up the droppings of infected fowls.

Moore says: "This disease of fowls has not been found in flocks where a good sanitary régime has been enforced. It is highly probable that it is a filth disease, being dependent upon unfavorable environments quite as much as the specific organism for the ability to run a rapidly fatal course and of spreading to the entire flock."

Treatment. — Prevention is the only known treatment. A maintenance of generally sanitary conditions and the avoidance of the introduction of diseased birds are effectual. If the disease appears in the flock separate the diseased birds, disinfect the premises, and place the flock under sanitary conditions. The disease will probably disappear, as it is difficult experimentally to maintain an infection when the birds are kept under sanitary conditions.

Prognosis. — Diseased birds usually die in from a few hours to two weeks, but they may recover.

The Sleepy Disease (*Apoplectiform septicæmia*)

This parasitic blood disease is apparently rare.

Diagnosis. — The most striking symptom is sleepiness.

According to Dammann and Manegold [1] the affected fowls show a roughness of plumage, swollen eyes, paleness of the comb and lameness. At autopsy symptoms of hemorrhagic septicæmia are found. The musculature is permeated with bloody effusions and red spots are observed in the mucous membrane of the intestines. The spleen is considerably enlarged and hemorrhagic patches were observed in other parts of the body.

Etiology. — According to Dammann and Manegold this disease is caused by a capsule bearing streptococcus (*Streptococcus capsulatus gallinarum*). The organism is present both in the blood and the infected organs. The disease may be readily transmitted by inoculation of virulent blood to other chickens. The incubation period varies from 6 to 14 days and the course of the disease from 1 to 3 weeks. It is not known how the disease is naturally transferred from one bird to another.

Treatment. — No treatment is known except the maintenance of general sanitary conditions.

Spirochætosis

This disease has not yet been reported in this country. It is known in South America, Europe, Africa and Australia. It may exist in this country undistinguished from fowl cholera.

Diagnosis. — There is a dullness, loss of appetite and thirst. The birds stand with head and tail down and eyes closed as in Fig. 34.

There is a rise of temperature. Diarrhea is present. There is a pronounced anæmia. Post-mortem examination

[1] Dammann, C., and Manegold, O., *Deut. Tierarztl. Wchnschr.* Vol. 13, pp. 577–579, 1905. And *Archiv Wiss. u. Prakt. Tierheilk.* Bd. 33, 41–70, 1907.

shows enlargement of liver and spleen. The crisis of the disease occurs on the fourth or fifth day. In fatal cases the fever disappears and the temperature sinks to below normal shortly before death.

Etiology. — This disease is caused by a spirochæte (*Spirochæta gallinarum*) found in the blood and in the liver and spleen. According to Balfour[1] the parasitic organisms

Fig. 34. — Bird suffering from spirochætosis. (From Kolle and Hetsch.)

enter the blood corpuscles. Both the organisms and the corpuscles then degenerate. According to Lounoy and Bruhl[2] the number of erythrocytes may be reduced one-half in five days. A favorable turn at the crisis of the disease is due, according to Levaditi and Manouclian,[3] to the

[1] Balfour, A. S., "Spirochætosis of Sudanese Fowls — an After Phase." *Jour. Trop. Med. and Hyg.*, Vol. 11, p. 37, 1908.

[2] Lounoy, L., and Bruhl, M. L., *Ann. Inst. Pasteur*, T. 28, pp. 517–539, 1914.

[3] Levaditi and Manouclian, *Ann. Inst. Pasteur*. T. 20, pp. 593–600, 1906.

destruction of the parasites by the large leucocytes of the spleen and liver. The organisms are carried from one fowl to another by the tick *Argas* (see p. 228). According to von Prowazek[1] the tick is a true intermediary host, the organisms appearing in the salivary glands about 14 days

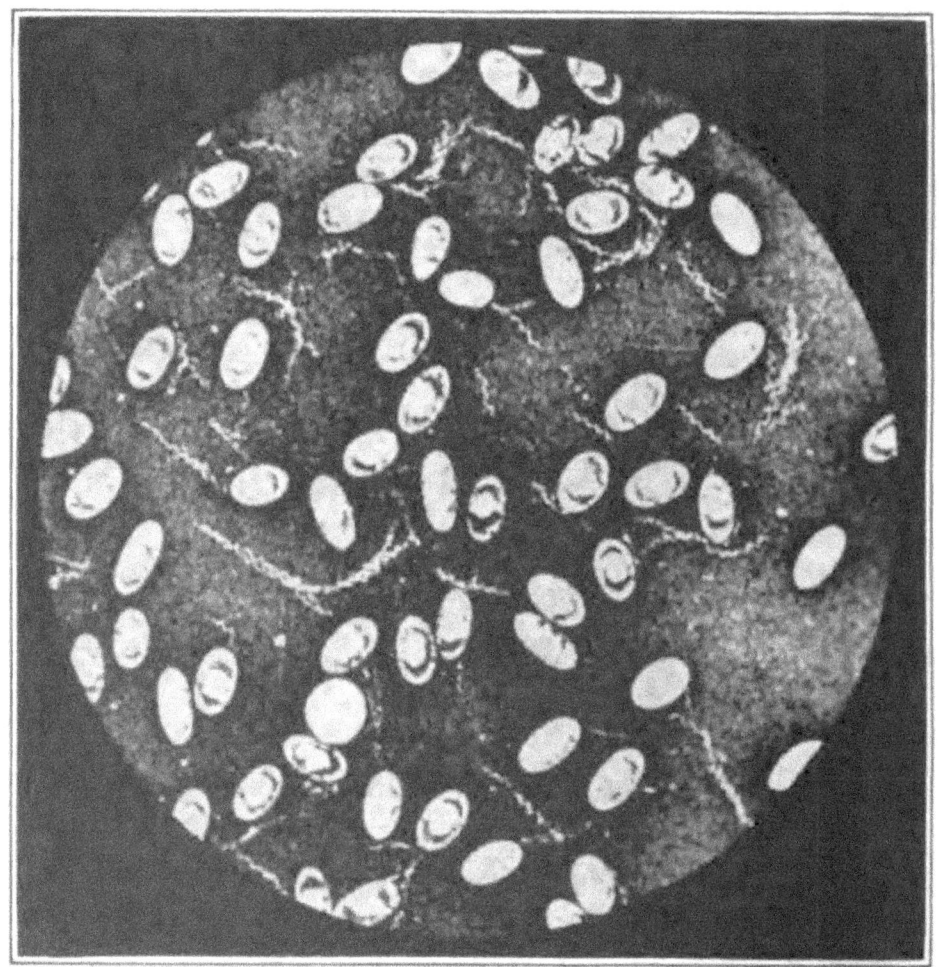

Fig. 35. — Fowl spirochætosis. (From Kolle and Ketch, after Burri.)

after infection. The organisms may live in the body of the tick for seven or eight months. That the tick is not a necessary host is shown by the fact that injection of the

[1] Von Prowazek, S., *Mem. Inst. Oswaldo Cruz.* T. I, pp. 79–80, 1911.

blood of an affected fowl can produce the disease in a healthy one.

Treatment. — The disease has been successfully treated with atoxyl,[1] and with salvarsan (606).[2] The best treatment is prevention by keeping the fowls free from ticks.

Prognosis. — About one-third of the affected fowls recover if untreated.

[1] Levaditi et McIntosh, "L'influence de l'atoxyl sur la spirillose provoquée par le Spirillum gallinarum." *Comptes Rendus Soc. Biol.*, T. 62, 1907.

Uhlenhuth u. Gross, "Untersuchungen über die Wirkung des Atoxyls auf die Spirillose der Hühner." *Arb. K. Gsndhtsamt.*, 15d, 27, pp. 231–255, 1907.

[2] Hauer, A., *Centbl. Bakt.* Bd. 62, pp. 477–496, 1912.

CHAPTER XIII

Diseases of the Nervous System

Apoplexy (Hemorrhage of the Brain)

In this disease the bird usually drops dead or paralyzed without showing any previous sign of illness. The only abnormality found on examination of the dead bird is clotted blood on the brain.

Etiology. — The cause of this disease is the rupture of a blood vessel in the brain and the pressure on the brain due to the blood which escapes. The cause of this rupture may be an unhealthy condition (usually a fatty degeneration) of the walls of the brain blood vessels. The immediate cause of the rupture is increased blood pressure due to fright, over-exertion, or strain in laying (hens often die on the nest). This disease is more apt to attack very fat birds and the degeneration of the vessels is supposed to be due to too rich food or to overfeeding.

Treatment. — Treatment of the affected birds is useless. So-called "apoplexy cures," of which there are some on the market, should be left strictly alone by the poultryman. Only *very rarely* can apoplexy be recognized till after the bird is dead, and then all the pills or potions ever invented for the purpose of swindling a gullible public will be of no avail. If *several successive* deaths from apoplexy occur, modify the ration, giving more green food and less meat and corn. See that the birds have plenty of range.

Prognosis. — The bird is usually found dead or dies in a little while.

Heat Prostrations

In very warm weather heat prostrations may occur, especially among heavy fowls. This is sometimes considered to be the same thing as apoplexy. The birds suddenly drop insensible or paralyzed.

Etiology. — The cause is pressure on the brain, due to heat, but the blood vessels are not ruptured as in apoplexy.

Treatment. — Mild cases may be treated by applying cold water to the head and keeping the bird in a cool, quiet place.

Prognosis. — Mild cases may recover. Others usually result fatally in a short time. As a preventive avoid overcrowding in hot weather. If the range is not provided with natural shade, supply artificially shaded places in which the birds may find protection from the hot sun during the middle of the day.

Congestion of the Brain (Vertigo, Cerebral Hyperæmia)

A number of abnormal physiological conditions may lead to a congestion of blood in the brain. This is usually associated with a diseased condition of other organs, and hence often occurs as a complication with other diseases. It is sometimes due to injury of the head.

Diagnosis. — Pearson and Warren [1] give the following diagnosis of this disease: "It is characterized by staggering, stupor, unusual movements such as walking backward or walking in a circle, unusual and irregular movements with the wings and feet and twisting the head backward or to the side. Sometimes the bird will fall on its side and make peculiar movements with its feet and wings as though attempting to run or fly."

[1] Pearson and Warren, "Diseases and Enemies of Poultry." 1897.

Etiology. — The congestion of the brain is sometimes due to blows on the head or to fright or other intense excitement. Often it is associated with acute indigestion or with the presence of parasitic intestinal worms.

Treatment. — Apply cold water to the head. Administer a laxative (2 teaspoonfuls of castor oil, or 30 grains of Epsom salts given in water or $1\frac{1}{2}$ grains of calomel). Keep the fowl in a cool, quiet place. If this treatment is not efficient Salmon recommends 1 to 5 grains of bromide of potassium dissolved in 1 tablespoonful of water 3 times a day. If intestinal worms are found in the droppings after the laxative, treat for the removal of these parasites (p. 139).

Prognosis. — The bird may recover if the cause is removed.

Epilepsy

This somewhat rare disease is characterized by occasional fits. Between these the birds appear normal.

Diagnosis. — Pearson[1] describes the behavior of the bird during the fit as follows: "The fowl will make beating movements with its wings, its legs will draw up and it will fall down, sometimes turn over on its back, or it may stand upright with its legs apart, head turned backward and mouth and eyes opening and closing spasmodically."

This spasm passes away after a time and leaves the bird in a normal condition.

Etiology. — It is often impossible to discover any cause of the disease. It is said to be sometimes caused by tumors on the brain and sometimes by intestinal worms.

Treatment. — The only cases that can be treated are those caused by the presence of intestinal worms. An affected bird should be put up and given a laxative and if intestinal

[1] *Loc. cit.*

worms are passed treat the patient for the removal of these parasites (p. 139).

The birds may live some time with occasional fits and may recover. Cases caused by intestinal worms are definitely cured by removing the parasites.

Polyneuritis, or Beri-beri

A nervous disease of fowls resembling human beri-beri is known in India, the Philippine Islands and Europe.

Diagnosis. — The chief symptom is a progressive paralysis of the legs. The nerves supplying the affected parts are greatly changed, often showing an almost complete disappearance of nerve fibers.

Etiology. — It has been known for several years that this disease occurs when the diet of chicks or fowls is completely or nearly completely confined to rice or other cereals from which the outer coat has been removed. If the whole grain is fed the disease does not occur. It has been shown that this is a true deficiency disease caused by a diet which lacks some substance which is essential for the normal metabolism of nervous tissue. The addition of milk, meat, legumes, rice polishings, or potatoes to a deficient diet prevents the disease. During the last two years Funk[1] and others have studied the nature of the substance or substances which must be added to the deficient food. They have isolated from rice polishings a crystalline alkaloid designated as Funk's base or vitamine. According to Vedder and Williams[2] this "probably exists in the food as a pyrimidin base combined as a constituent of

---

[1] Funk, C., "Die Vitamine." Wiesbaden, 1914.
[2] Vedder, E. B., and Williams, R. R., "Concerning the Beri-beri Preventing Substances or Vitamins Contained in Rice Polishings. A Sixth Contribution to the Etiology of Beri-beri." *Philippine Jour. Sci.*, Sect. B., Vol. 8, pp. 175–195.

nucleic acid but that it is not present in the nucleins of nucleic acids that have been isolated by processes involving the use of alkalies or heat."

Treatment. — Prevention is the best treatment. This disease can be easily prevented by feeding a rational diet. The disease may be cured by feeding rice polishings or an extract made from them. It can also be cured by a dose of 30 mg. of Funk's base. This is, of course, not available for therapeutic purposes.

Prognosis. — If the cause is not removed the birds die. If the food is changed so that it has a sufficient supply of the deficient material they recover.

CHAPTER XIV

DISEASES OF KIDNEYS, RHEUMATISM AND LIMBERNECK

In routine autopsy work where all dead birds are examined probably no organ except the liver is more frequently found in a diseased condition than the kidneys. They are often enlarged. Sometimes they contain dark points caused by the rupture of small blood vessels, and in other cases they may contain abscesses. Micro-organisms have been obtained from some cases of diseased kidneys. Nothing is yet known of the causes of these specific diseased conditions in poultry. Some of the cases of under-development, especially of pullets, are apparently due to enlarged kidneys. In such cases the birds usually lose their appetite, become emaciated and their feathers are roughened. No dependable diagnosis of diseased kidneys can be made on the living fowl. When several cases occur care should be taken to see that the flock receives a balanced ration with plenty of green food, as diseased kidneys may occur from too much protein in the food.

One of the diseased conditions of the kidneys results in an inability to eliminate the urates. The uric acid content of the blood is greatly increased and the urates are deposited on the surface of the visceral organs, in the tissues of the urinary apparatus and around the joints in the form of crystals or urate of soda.

Gout

This diseased condition is called gout. In fowls as in man it has two forms, the visceral and the articular, depending upon the location of the deposits of urates.

Visceral Gout

In visceral gout the only symptoms shown by an affected bird are a loss in weight or "going light" and a slight yellowish tinge to the skin, comb and wattles. The bird has a good and often abnormal appetite. Death occurs suddenly. An examination of the abdominal cavity shows that all the organs and serous membranes are covered with a chalky or talcum-like powder. This powder has a mother-of-pearl luster and on microscopic examination is seen to be composed of small needle-like crystals. These are crystals of urate of soda. These crystals are also found in the urinary organs. The ureter and collecting tubules are often filled with a mass of these crystals. Hebrant and Antoine give the following test for the urate of soda.

Dissolve the crystals in nitric acid and evaporate in a watch glass. This gives a red onion peel mass which turns purplish blue on the addition of a solution of caustic potash.

Articular Gout

In this form of the disease the crystals or urate of soda are in nodules around the joints, especially of the feet and toes. These nodules sometimes appear like strings of beads on the under side of the toes. They contain a white or creamy thick liquid composed mostly of the crystals. They are at first soft but later become very firm. The presence of the nodules causes stiffness and soreness of the joints and the birds become indisposed to stand or walk. Sometimes the nodules ulcerate, discharging a stringy pus and exposing the cavities of the joints to the air. The development of fistulas causes the death of the bones. The disease is slow in its development and advanced stages are seen only in old birds. The birds lose weight and in advanced stages diarrhea sets in and death from exhaustion follows.

Early stages of this disease are often mistaken for rheumatism on account of the stiffness and soreness of the joints.

Etiology. — The cause of this disease is a disturbance of the normal physiology of excretion so that the uric acid which should be excreted by the kidneys is first retained in the blood and then deposited within the body as crystals of urate of soda. The disturbance is probably due to a diet which is too rich in proteids. It has been experimentally produced by feeding meat.[1] Beef liver produces the condition more quickly than horse meat.

Treatment. — In case of articular gout Salmon recommends rubbing the affected joints with camphorated or carbolic ointment. In well developed cases it is more profitable to kill the birds than to treat them. Visceral gout is not usually recognized while the bird is alive. Prevention is the only reliable treatment for either form of gout. Birds should be kept under sanitary conditions and given plenty of green food. When several birds develop the disease it is well to give the whole flock Epsom salts ($\frac{1}{2}$ to 1 teaspoonful per bird) and to reduce the amount of meat scrap and increase the quantity of green food.

Prognosis. — The disease, especially the articular form, is chronic and advanced cases are only found in old birds. Badly diseased birds may live a long time. Mild cases may recover on corrected diet.

Rheumatism

A lameness or stiffness is usually considered rheumatism. Many such cases are due to tuberculosis of the joints (p. 121), and others to articular gout, but there are muscular and joint inflammations caused by exposure which are properly con-

---

[1] di Gristiana, G., *Internat. Beitr. Path. u. Ther. Ernährungsstör. Stoffw., Verdauungskrank.* Bd. 1, pp. 29–47.

sidered rheumatism. This disease is an inflammation of the connective tissues of the muscles and joints.

Etiology. — It is caused by exposure to cold or dampness. The occurrence of several cases in the flock indicates something wrong in the housing conditions.

Treatment. — The disease is prevented by keeping the fowls in dry, warm, well ventilated houses with well drained runs.

Prognosis. — Fowls protected from further exposure and given a good ration with plenty of green food usually recover.

Limberneck

This is not properly a disease but a *symptom* which accompanies several diseased conditions. A fowl is said to have limberneck when partial or entire nervous control of the neck muscles is lost. The neck may hang limp so that the head falls on the ground between the feet. Sometimes the bird is able to raise the head from the ground by making a great effort.

A bird is sometimes said to have limberneck when the dorsal or lateral neck muscles are tense, the head drawn convulsively backward, but this is more often called "wry-neck."

Both limberneck and wry-neck are due to nervous disorders which arise from several different causes. "Wry-neck" is usually associated with direct brain or nerve irritation and occurs in epileptic spasms, but also sometimes is associated with rheumatism. Limberneck is usually associated with colic, acute indigestion, intestinal parasites, or ptomaine poisoning.

No treatment for limberneck *as such* can be advised. Effort should be made to ascertain and cure the diseased condition which is responsible for this symptom.

Cases due to rheumatism, colic, indigestion, intestinal parasites, and some of those due to poisoning may recover, if the real cause can be ascertained and treated soon enough.

CHAPTER XV

External Parasites

VIGILANT and continuous attention is necessary to keep fowls free from external parasites. At least 32 species of arachnids and insects are known to be parasitic on fowls. Some of these like the red mites visit their host only to take food and spend the rest of the time on the under side of the roosts, in cracks and crevices and various other places of seclusion. Others like the lice normally stay on the birds, although occasionally some individuals crawl off, especially into the nest. Some of these parasites live upon the surface of the skin and upon the feathers, deriving their nourishment either by sucking the blood like the red mite, or by chewing the skin and feathers like the lice and some of the mites. Some of the mites, however, bore under the skin, causing skin diseases known as scabies or psoric diseases. The most common of these diseases are scabby or scaly leg and depluming scabies.

The economic importance of these external parasites is very great. Fowls infested with one or several of these species of parasites are not profitable. They make a smaller growth in the same time with the same food and their egg production is not equal to similar birds not so infested. Not only are they constantly robbed of some of their tissue and blood but their rest is disturbed. Sleep is as important to the normal physiology of a bird as it is to that of a man.

Keeping a Poultry Plant Free from External Parasites. — It is not necessary for a poultryman to be able to distinguish the 32 species of parasites or to know their life histories in order to keep his plant free from them. It is only necessary to know that some of them stay on the birds and can only be exterminated by treating the birds, while others spend most of their time on the under sides of the roosts in cracks and can best be exterminated by contact sprays containing cresol or kerosene. A single application is not efficient in either case but treatment must be repeated 2 or 3 times at intervals of a few days to destroy those that hatch after the treatment or are concealed beyond its reach. A routine procedure by which a poultry plant can be kept free from parasites is very useful. The following method has proven very successful at the Maine Experiment Station.

All hatching and rearing of chickens is done in incubators and brooders. The growing chickens are never allowed to come into any contact whatever with old hens. Therefore, when the pullets are ready to go into the laying houses in the fall they are free from lice. Sometime in the later summer, usually in August or early in September, the laying houses are given a thorough cleaning. They are first scraped, scoured and washed out with water thrown on the walls and floor with as much pressure as possible from a hose. They are then given two thorough sprayings, with an interval of several days intervening, with a solution of cresol such as is described in Chapter II. Then the roosting boards, nests, floors and walls to a height of about 5 feet are thoroughly sprayed with the lice paint (kerosene oil and crude carbolic acid described on page 15).

For ridding the birds of lice the Maine Station formerly recommended dusting two or three times at intervals of several days to a week with the lice powder described on page 211. All birds which were to be kept over for the next year's work were treated in this manner before they were put into the cleaned houses.

During the past few years this Station has adopted another method of freeing the birds of lice, namely the use of mercurial ointment. An ointment has certain very distinct advantages over any powder. It is much easier to apply, and requires less of the poultryman's time, which is an important factor on a large plant. Further a mercurial ointment is more efficient and lasting in its effect as a parasiticide than any powder.

The form of ointment which we have used is the ammoniated mercurial ointment U.S.P., with the exception that we have the druggist make it with a lard base, instead of the lanolin base called for by the Pharmacopœia. Others use the blue ointment of the U.S.P. for the same purpose. It is probably more effective, part for part, but it is also more expensive.

The proper method of applying the ointment is to rub well on to the skin three pieces of the ointment, each as big as a small pea. One of these pieces should be rubbed on just under the vent, the other two under the wings. The ointment should never be simply daubed on and left as a lump. If it is so done, the bird is very liable to get some of it in the mouth and a case of mercurial poisoning will result.

As a result of these methods the Station's poultry plant is at all times of the year practically free of lice.

This method keeps the flock free from lice and the mites which live upon the surface of the skin, but would not destroy those mites which penetrate the skin and cause scabies. These and other more rare parasites should be destroyed when present by special methods. The description of, and treatment for, each class of external parasite is given below.

A. LICE (MALLOPHAGA)

Lice are probably the most widely distributed parasite of poultry. They are so common that flocks of fowls that have not been treated to remove lice for a long time are almost sure to have one or more species present. At least 8 species of hen lice have been found and 5 of these are common. Bird lice are quite different from those which affect man and mammals.

The popular notion that lice may be transmitted from poultry to other animals is quite erroneous. Theobald says: "So particular are bird-lice that it is quite the exception to find one species upon two distinct kinds of birds. Fowl-lice will not even attack the duck nor duck-lice the fowl. Nearly every bird has its own particular Mallophagan parasite or parasites. They may possibly pass to some strange host for a short time, but they will not live and breed. Moreover, ... particular species attack restricted areas on the same host and are seldom found in other positions." Some of these lice are sluggish, nearly stationary, and confined to a restricted area of the body, while others are active and crawl over the entire body. Theobald describes eight species of lice found on poultry.

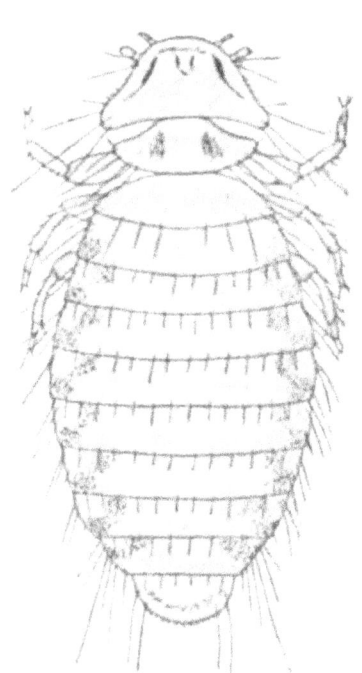

FIG. 36. — The common hen louse (*Menopon pallidum*). Greatly enlarged. (From Banks.)

The most common and widely distributed hen louse found in this country is *Menopon pallidum*. This louse is shown in Fig. 36.

Another species of this genus (*Menopon biseriatum*), which

closely resembles *M. pallidum*, is also sometimes found. These are active lice living on all parts of the body. They often crawl on to the hands when handling or plucking birds, and may sometimes be found in the nests.

There are several other lice which sometimes infest poultry. Each of these species is confined to a special region of the host. Although capable of crawling about, the lice of these species for the most part remain nearly stationary, often with their heads buried in the skin and their bodies erect. Two species, *Lipeurus variabilis* and *Lipeurus heterographus*, live among the barbs of the wing and tail feathers. *Goniodes dissimilis* is found under the wings and on the rump. The appearance of two of the species mentioned, viz., *Lipeurus variabilis* and *Goniodes dissimilis*, is shown in Figs. 37 and 38 respectively.

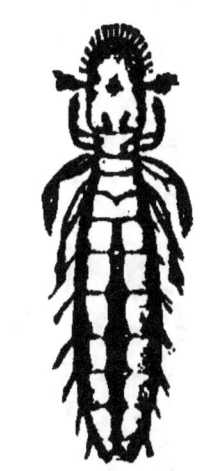

Fig. 37.—*Lipeurus variabilis*. A louse that infests poultry. Much enlarged. (From Banks, after Denny.)

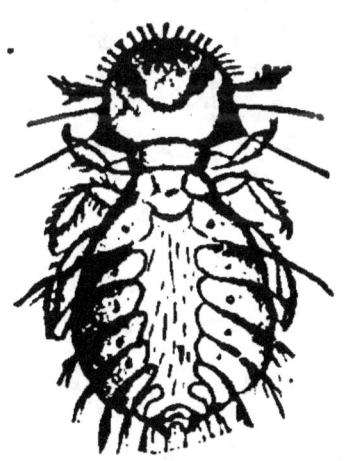

Fig. 38.—*Goniodes dissimilis*, a louse that infests poultry. Much enlarged. (From Banks, after Denny.)

All true bird lice (*Mallophaga*) have biting, not piercing mouth parts. They live upon fragments of feathers, scales of the skin and other such débris. It is evident from the action of infested birds that these parasites cause considerable pain and itching. This must be particularly true when they are present in large numbers.

Life-history of Lice.— All the lice breed fairly rapidly. The eggs or nits are laid upon the down feathers, as a rule; they are often beautifully sculptured objects, oval in form. In about 6 to 10 days they hatch into small, pale, active lice, which at once commence to irritate the birds. The adults are occasionally found in the nests. Some species are found copulating in the nests, others always on the birds. They live a considerable time. *Menopon pallidum* (Fig. 36) has been kept alive for months upon fresh feathers, the quill epidermis being especially eaten. Before reaching the

Fig. 39. — Feathers showing eggs or "nits" of the common hen louse. Enlarged. (Original.)

full-grown state as many as 10 or 12 molts apparently take place, there being little difference in each stage, except the gradual darkening of the markings.

The eggs or nits of hen lice are shown in Fig. 39.

Methods of Introduction and Infestation. — It is generally agreed that lice and other parasites flourish best in insanitary surroundings. There must, however, be a source of infestation. Lice are brought to a new place by introducing infested birds. They spread from bird to bird (a) directly during copulation (an infested cock often infests the whole flock), or (b) when two hens occupy a nest together, or (c) from mother to chick. They also pass indirectly from bird to bird by crawling off one bird first on to the nesting material and later on to another bird which uses the same nest. Sharp has also observed several lice clinging to the body of a fly parasitic upon chickens. Lice are so much more common than the parasitic fly that it is probable that this insect is of little real importance in the distribution of the lice.

All the lice breed very rapidly. In 8 weeks the third generation is mature and in this generation the estimated number of the offspring of a single pair is 125,000 individuals. It seems important to eradicate an infestation if possible as soon as discovered. However, if kept under sanitary conditions and furnished with plenty of attractive dust, vigorous birds will hold external parasites in check. With some attention to sick birds, setting hens and young chicks, the parasites will give little trouble on a plant conducted with due regard to the principles of hygiene and sanitation (cf. Chapter II).

Salmon ("Diseases of Poultry") says: "It should be remembered at all times that the external animal parasites are the most common and frequent cause of trouble in the poultry-yard and pigeon-cote. If the birds are not thriving and conducting themselves satisfactorily, look for these pests, take measures to repress them, and in most cases the results will be surprising and gratifying. When anything is the matter with a horse the maxim is *examine his feet*, and when anything is found wrong with poultry or other domesticated birds, the maxim should be *look for lice.*"

Diagnosis. — Adult hens may harbor quite a number of these parasites without showing any symptoms which indicate their presence. If they are unthrifty and broody hens leave their nests they should be examined for lice. The biting and digging of the claws of the lice may cause sores and the nervous irritation and loss of sleep may cause general debility and bowel trouble. Little chickens are very susceptible and often die. Lice are frequently found in large numbers on birds suffering from roup, gapes, etc. In some cases their presence has rendered the birds more susceptible to other disease, while in others it is probable that the birds lack sufficient energy to dust themselves.

The sure test for the presence of lice is, of course, finding the lice. Part the feathers under the wing, on the back and around the vent and examine the exposed skin. Examine the head and neck feathers and look between the large feathers of the wing. When present the parasites are easily found by any one who is familiar with them. It seems incredible that serious infestations can escape the eye of any poultryman.

Treatment. — Sanitary surroundings and liberal range help the birds in their attempts to keep themselves free from lice. The dust bath is very efficient in holding the pests in check. It is doubtful, however, whether the dust boxes which used to be almost universally kept in the poultry house are of any real value. It is a noticeable fact that dust boxes are much less used now than formerly. As commonly made these boxes are too small, and too shallow, and are not filled with the proper kind of material. Hens will use them, in most cases, only as a last resort if at all.

When possible, birds should be given access to dry, sandy ground, and they will provide their own dust bath. Some authors advise adding insect powder to the earth in dust boxes for bad infestations. It is doubtful whether under the best of circumstances this does anything but waste the insect powder. It is better to apply the powder directly to the bird and furnish clean earth for the dust bath.

When hens are used for incubating and brooding it is

necessary to give some individual treatment to brooding hens and young chicks. It is also necessary to treat sick hens which are not able to use the dust bath. While it is theoretically possible to exterminate the pests and keep the flock free from them by avoiding the introduction of infected birds, this ideal condition prevails in very few poultry plants. In almost all flocks there are enough lice present to cause trouble if conditions favor their development.

How to Make an Effective and Very Cheap Lice Powder.— When the treatment of individual birds for lice becomes necessary some kind of powder dusted into the feathers thoroughly has been one of the most effective and advisable remedies. The powder used must be of such nature, however, that it will be *effective*. There are so-called "lice powders" on the market which are no more effective than an equal quantity of any inert powdered substance would be. It is not only a waste of money but of time as well to use such powders. At the Maine Station no lice powder has been found that is so satisfactory as that originally invented by Mr. R. C. Lawry, formerly of the poultry department of Cornell University. The following matter regarding this powder (which can be made at a very low cost) is quoted from a circular issued by the Maine Station:

"In using any kind of lice powder on poultry, whether the one described in this circular or some other, it should always be remembered that a single application of powder is not sufficient. When there are lice present on a bird there are always unhatched eggs of lice ('nits') present too. The proper procedure is to follow up a first application of powder with a second at an interval of 4 days to a week. If the birds are badly infested at the beginning it may be necessary to make still a third application.

"The lice powder which the Station uses is made at a cost of only a few cents a pound in the following way:

"*Take 3 parts of gasoline,*
    1 part of crude carbolic acid;

"To get the proper results *only the 90–95 per cent carbolic acid should be used for making lice powder.* Weaker acids are ineffective.

"Owing to the difficulty in getting the strong crude carbolic acid locally in this State at reasonable prices, the Station has experimented to see whether some other more readily obtainable substance could not be substituted for it. It has been found that *cresol* gives as good results as the highest grade crude carbolic.

"The directions for making the powder are now, therefore, modified as follows:

"Take *3 parts of gasoline, and*
    1 part of crude carbolic acid, 90–95 per cent strength,
or, *if the 90–95 per cent strength crude carbolic acid cannot be obtained* take
    3 parts of gasoline and
    1 part of cresol.

"Mix these together and add gradually with stirring, enough plaster of paris to take up all the moisture. As a general rule it will take about 4 quarts of plaster of paris to 1 quart of the liquid. The exact amount, however, must be determined by the condition of the powder in each case. The liquid and dry plaster should be thoroughly mixed and stirred so that the liquid will be uniformly distributed through the mass of plaster. When enough plaster has been added the resulting mixture should be a dry, pinkish brown powder having a fairly strong carbolic odor and a rather less pronounced gasoline odor.

"Do not use more plaster in mixing than is necessary to blot up the liquid. This powder is to be worked into the feathers of the birds affected with vermin. The bulk of the application should be in the fluff around the vent and on the ventral side of the body and in the fluff under the wings. Its efficiency, which is greater than that of any other lice powder known to the writer, can be very easily demonstrated by any one to his own satisfaction. Take a bird that is covered with lice and apply the powder in the manner just described. After a lapse of about a minute, shake the bird, loosening its feathers with the fingers at the same time, over a clean piece of paper. Dead and dying lice will drop on the paper in great numbers. Any one who will try this experiment will have no further doubt of the wonderful efficiency and value of this powder."

Next to the Lawry powder probably pure pyrethrum or Persian insect powder is as cheap and effective as anything to be had.

A time-honored and effective treatment for lice, especially for young chicks, is greasing. The grease most often used is lard or sometimes lard and sulphur. The latter should not be used for young chicks. The lard is applied with the finger to the head, neck, under the wings and around the vent. Greasing is a somewhat tedious but very effective treatment for lice, especially on young chicks, since lice usually attack them on the head and neck.

B. MITES — ACARINA

Eighteen species of mites are parasitic upon fowls. Only 4 of these are sufficiently injurious and widely distributed to be of great economic importance. Occasionally one or another of the other species becomes sufficiently abundant to be of local importance. The mites are small 8-legged animals related to the spiders. Some of the mites parasitic on the fowl visit their host only to feed, as the common red mite; others remain on the surface of the skin or on the feathers, as in the case of depluming scabies. Others live under the skin, causing deep-seated skin diseases like scaly leg; still others find their way into the internal regions of the body, living either on mucous membrances like the air-sac mite (p. 180) or upon the connective tissue like the connective tissue mite.

The most common and most injurious mite parasitic on fowls is the common fowl mite or red mite, *Dermanyssus gallinæ*. These mites are present in almost every poultry house that is not kept very clean. When they are present

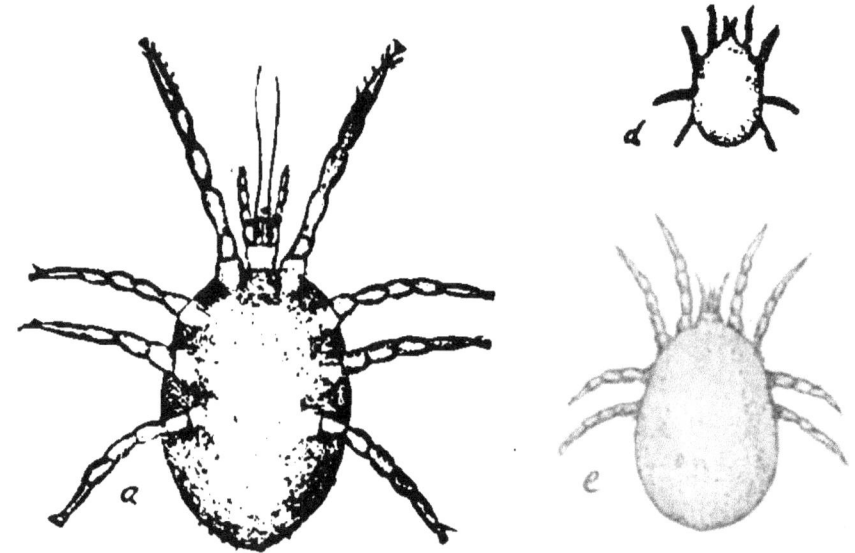

Fig. 40. — The common "red mite" of poultry, *Dermanyssus gallinæ*. a, adult. d and e, young. (After Osborn.)

in large numbers they are a serious pest. This mite is a little more than ½ millimeter long. The female is a little larger than the male. When empty they are gray with dark spots, but usually they appear some shade from yellow to dark red according to the amount of fowl's blood they contain. They visit the fowls only to feed and spend the rest of the time on the under sides of the roosts, in cracks and crevices, under collections of droppings or other filth and in the nesting material, especially if such material is dirty straw. The mites breed in these places. They reproduce very rapidly, especially in spring and summer. The eggs are laid in concealed places, usually in cracks containing filth or in dirty nesting material. The young mites are white and have only 6 legs. Their first food is probably filth or decayed wood. They molt several times and their cast skins are often seen as a white powder on the perches. After the first molt the larvæ have 8 legs. The mites are able to live and reproduce for months at least without animal food, but when they are associated with fowls the older larvæ and adults depend upon the blood of the fowls for food. They usually attack the birds at night but sometimes are found feeding on laying or brooding hens during the day. They pierce the skin with their needle-like jaws and suck the blood. The irritation due to the biting of a number of these creatures disturbs the rest of the bird and the loss of blood may be considerable. The mites thrive best in dark, damp, dirty houses and may be found in such houses for months after all fowls have been removed. They will bite man or other mammals, causing severe irritation, but do not remain on strange hosts for any length of time. Fowls should not be allowed to roost in sheds with other animals, as the sheds may become infested with the mites which will disturb the other animals as well as the fowls.

Diagnosis. — If the birds are not doing well, especially if they appear emaciated and dejected, they should be examined at night for mites. In the daytime the ends and under sides of the roosts and the cracks in them should be examined. Numbers of the mites are often found by prying up a loose cleat or splitting off a wide loose sliver. They may often be found in old straw nests.

Treatment. — Clean, dry, well ventilated houses which get plenty of sunlight are seldom badly infested. The first step in eradicating or controlling the pest is thoroughly to clean the houses. Remove the droppings and all the old nesting material. Clean and when possible scrub or wash with a stream from the hose all the perches, nests, floors and walls. Spray or paint the perches, nests, walls and floors with a 5 per cent solution of cresol (see Chapter II for directions for making this). Professor H. C. Pierce has tested various remedies for mites and finds none so effective as this. Use plenty of solution and make the spraying thorough. Every crack and crevice should be flooded.

Another spray successfully used is: 3 parts kerosene and 1 part crude carbolic acid. Still a third, kerosene emulsion is recommended by the United States Department of Agriculture. Their method of making this spray as given in Circular No. 92 is as follows: "To make this, shave $\frac{1}{2}$ pound of hard soap into 1 gallon of soft water and boil the mixture until the soap is dissolved. Then remove it to a safe distance from the fire and stir into it at once, while still hot, 2 gallons of kerosene or coal oil. The result is a thick, creamy emulsion. Dilute this stock mixture with 10 parts of soft water, and apply as a spray or with a brush, being careful to work it into all cracks, crevices, and joints of the building."

With any of these sprays it is necessary to make two or more applications at intervals of a few days to destroy the mites which hatch after the first application. The liquid may be put on with a hand spray pump or with a brush. Cleanliness, fresh air and sunlight are cheap and effective preventives.

Scaly Leg

A minute mite, *Knemidocoptes (Dermatoryctes) (Sarcoptes) mutans*, is the cause of a contagious disease affecting the legs of fowls, turkeys, pheasants, partridges and cage birds. According to some authorities it sometimes affects the comb and beak also. The mites excavate places under the skin

External Parasites

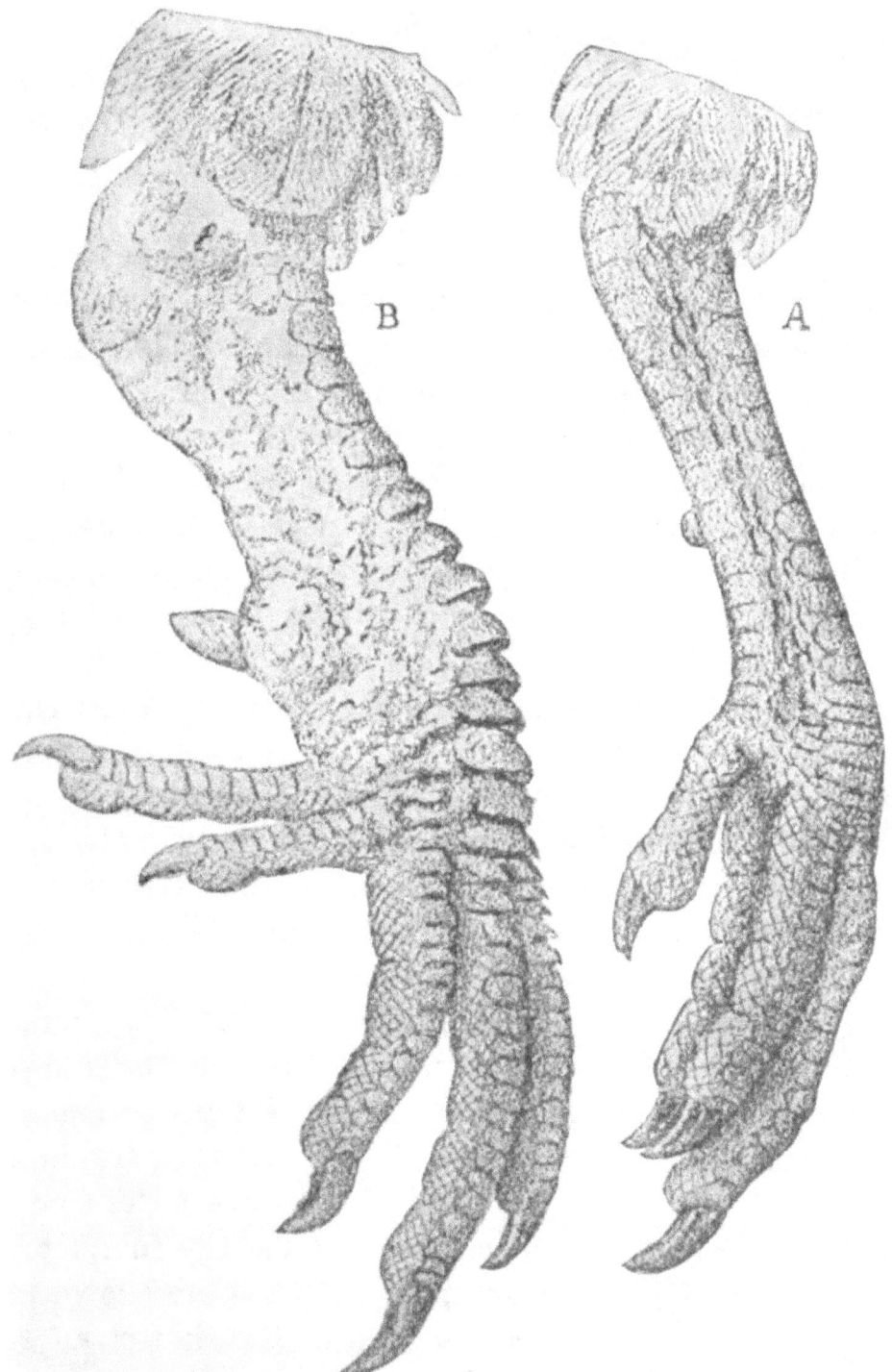

Fig. 41. — *A.* Normal leg of hen. *B.* Leg of hen affected with scaly leg. (After Mégnin.)

where they live and breed. The most thorough study yet made of this parasite and its effect on birds is that of Haiduk.[1]

Diagnosis. — This very common disease is easily recognized by the enlarged roughened appearance it gives the foot and shank. This appearance is shown in Fig. 41, with a normal leg for comparison.

The disease is present in most flocks unless especial care has been taken to exclude it. It is slightly contagious, but usually only a few birds in a flock appear to be infected. The scales on the foot and leg of an affected bird are raised by a crusty substance deposited beneath them. The lesions usually appear first near the joints between the toes and foot. The parts affected first appear to be enlarged and then the scales are raised, giving the roughened appearance shown in *B*, Fig. 41. In early stages the disease does not appear to disturb the general health of the fowl. As it progresses the birds become lame and sometimes the foot becomes so badly diseased that joints or even whole toes drop off. The photograph of a badly affected leg is shown in Fig. 42. The two legs are usually affected equally.

Etiology. — The disease is caused by the minute parasitic mite *Knemidocoptes mutans* (Figs. 43 and 44).

The mites bore under the scales of the foot and leg and burrow deeper and deeper into the tissue. They set up an irritation which leads to multiplication of cells and the exudation of serum. This accumulation forms crusty deposits beneath the scales. These crusts contain many depressions in which are embedded female mites containing eggs. The larvæ and the males are usually found beneath the crusts. The relations just described are shown in Fig. 45 which is a picture of a section of the skin of a "scaly" leg.

[1] Haiduk, T., "Die Fussräuder des Geflügels." *Inaug. Diss. Giessen*, 1909, pp. 1–58, Taf. I–VI.

Fig. 42. — Photograph of the leg of a hen very severely affected with scaly leg. (After Haiduk.)

As the disease progresses the mites which are becoming constantly more numerous penetrate very deep into the

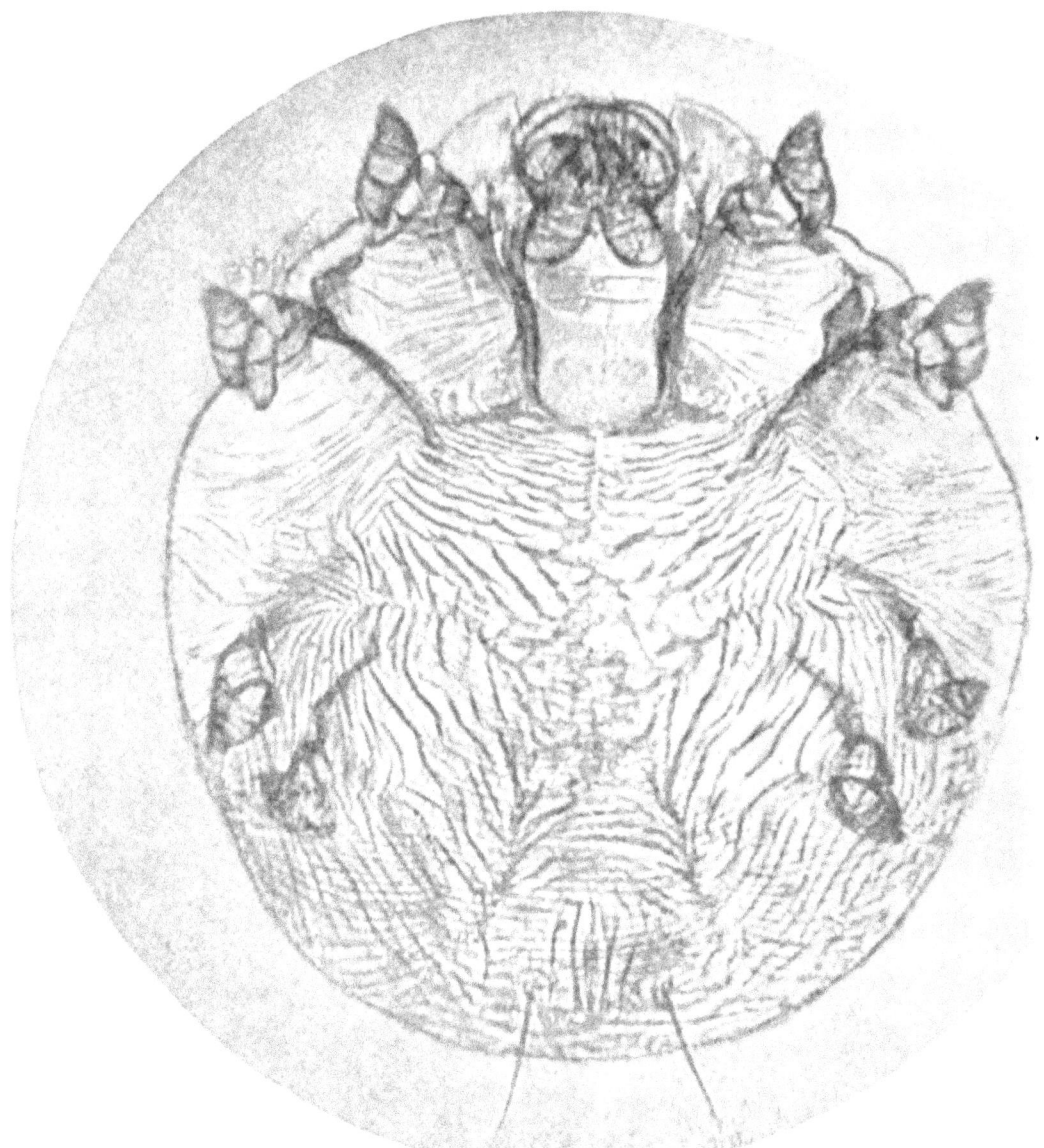

Fig. 43. — Photograph of the adult female of the mite *Knemidocoptes (Dermatoryctes) mutans*. (After Haiduk.)

tissues, causing lameness and sometimes the loss of some of the toes.

The infection from bird to bird probably takes place on the roosts or from mother to chick. Robinson believes that the birds most likely to be infected are those with a deficient supply of oil in the skin. The conditions which favor its

Fig. 44. — Photograph of the six-legged larva of *Knemidocoptes* (*Dermatoryctes*) *mutans*. (After Haiduk.)

spread in a flock are dry, barren runs, especially on alkaline soils or in yards filled with ashes or cinders. Foul roosting places also favor the spread of the disease. The disease is easily cured and it is worth the trouble of any poultryman to cure all the affected birds and to examine any birds purchased that infected ones may be treated before they are introduced into the flock.

Treatment. — Individual treatment is necessary to cure the disease. This treatment consists in the application of some penetrating oil to the diseased parts. A large number of oils and ointments have been used successfully. If the case is not far advanced and if there is no special hurry about bringing about the cure the application of the oils or ointments at intervals of 2 or 3 days will soon do the work. If the birds must be cured quickly for show or sale purposes the cure is hastened by removing the scales and crusts before applying the medicine. This may be done by brushing with a stiff toothbrush before each treatment. Or the feet may be soaked for a few moments in warm soapy water and then brushed. When the disease is far advanced it is best to begin the treatment by the removal of the scales.

Haiduk's experiments show that one of the very best cures for scaly leg is *oil of caraway*. *This is best applied in an ointment made of 1 part of oil of caraway to 5 parts of white vaseline.* Oil of caraway is very penetrating and is not nearly as irritating as some of the treatments more usually advised. This ointment should be rubbed into the leg and foot every few days until signs of the disease disappear.

Hill recommends daily application of an ointment made of equal parts of vaseline and zinc ointment, or in severe cases of one made of 1 ounce of sulphur, $\frac{1}{2}$ ounce of oxide of zinc, 1 dram of oil of tar and 2 ounces of whale oil mixed together.

There are two common remedies used successfully by poultrymen. These are irritating and should be used with some caution. They have the advantage of being quickly applied. The best of these is probably a mixture of 1 part

External Parasites 223

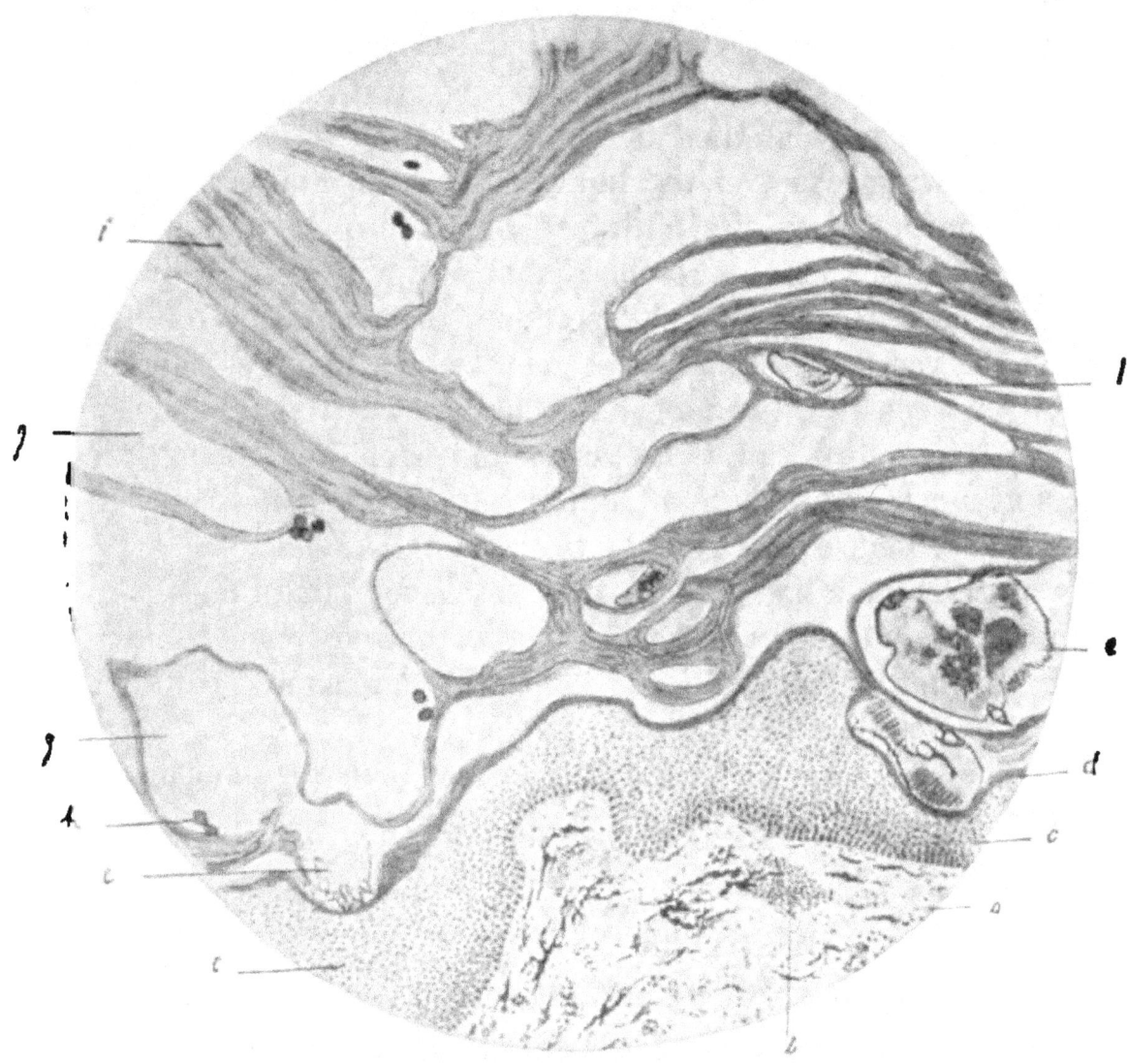

FIG. 45. — Section of the skin of the leg of a fowl affected with scaly leg.

    a. Papilla with pigment cells.
    b. Lymphatic tissue in the papilla.
    c. Epidermis: stratum profundum.
    d. Epidermis: stratum corneum.
    e. Section through a mite.
    è. Section through a mite showing head and 2 pairs of legs.
    f. Young mite.
    g. Cavity excavated by mites.
    h. Excrement of mite.
    i. Horny layer between the mite excavations.

(From Haiduk, after Olt.)

of coal oil or kerosene and 2 parts of raw linseed oil. If a quick cure is imperative a half-and-half mixture may be used. Robinson in *Farm Poultry*, May, 1907, recommends a quick and easy method of applying this. It is to take a tall quart measure of the liquid to the hen house at night and dip both legs of each infected bird into the measure of oil, holding them there for a moment and then allowing them to drip for a moment more and then replacing the hen on the roost. With any treatment which involves the use of kerosene care must be taken not to wet the feathers of the leg, as this causes irritation and sometimes burns the skin much as the human skin is burned when it is rubbed with kerosene and covered with flannel.

A second method of applying kerosene is to put a teaspoonful of the oil in a quart measure of water and treat the birds by the method given above. The same care should be taken not to wet the feathers.

The advantage of these treatments is their easy and rapid application to a number of birds.

Depluming Scabies

The mite *Sarcoptes lævis* var. *gallinæ* (Fig. 46) is the cause of a kind of scabies in fowls which causes the feathers to break off at the surface of the skin.

Symptoms. — This disease usually appears in spring and summer and is characterized by the dropping off of patches of feathers on different parts of the body. It usually begins at the rump and spreads to the head and neck, back, thighs and breast. The large wing and tail feathers are not usually lost. The exposed skin is normal in appearance. Around the stumps of the lost feathers and at the end of the quills of feathers near the bare spots are masses of epidermal scales. On microscopic examination these scales are found to be composed of numerous mites and their débris. The irritation of the mites often causes the birds to pull their own feathers. Birds affected often pull each others' feathers. Some of the so-called feather eating is due to the presence of this parasite, but fowls sometimes pull each others' feathers when the parasite is not present. Salmon says this disease does not affect the general health of the bird and does not appear to disturb gain in flesh or egg production, but Theobald says that the disease checks egg laying in hens and affected cocks become emaciated and sometimes die.

Etiology.—The mite *Sarcoptes lævis* which causes this disease is smaller than the one which causes scaly leg. They live at the base of the feathers in the epidermal débris referred to above. A flock becomes infested by the introduction of one or more birds carrying the mites. The mites are spread from bird to bird by the male in copulation. The distribution is often very rapid so that the whole flock is soon affected.

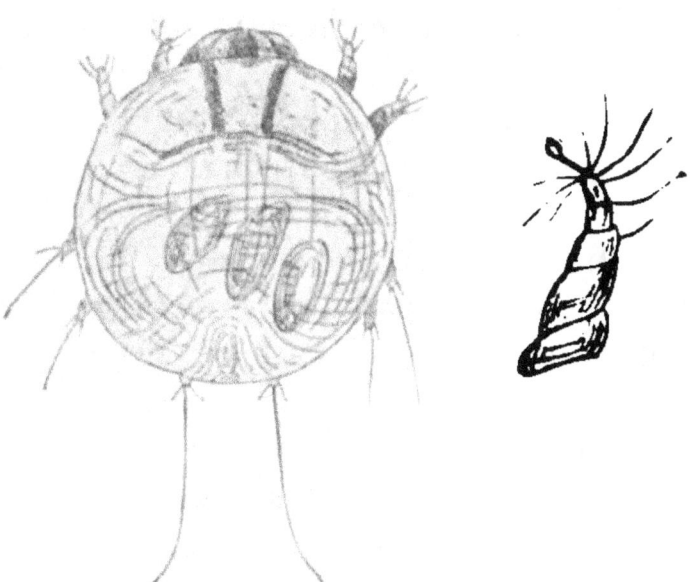

Fig. 46.— Egg containing female *Sarcoptes lævis* var. *gallinæ*. (After Theobald.)

Treatment. — The disease should be prevented by taking care not to introduce infested birds. If it appears, all affected birds should at once be isolated. The mites yield easily to treatment. The infested areas may be rubbed with some of the less irritating ointments recommended for scaly legs (see p. 222).

The following list gives some ointments in the order of their desirability for use on the body.

      Oil of caraway ointment (1 to 5).
      Balsam of Peru.
      Creolin treatment (1 to 10).
      Helmerich's ointment.

Salmon gives a modification of the latter ointment which he considers an improvement for use in depluming scabies.

      Flowers of sulphur, 1 dram,
      Carbonate of potash, 20 grains,
      Lard or vaseline, $\frac{1}{2}$ ounce.

Scabies may also be cured by liquid applications. The two following preparations are recommended by Salmon: A solution of balsam of Peru in alcohol (1 part of balsam to 3 of alcohol) or 1 dram of creolin, 2 ounces of glycerine, $\frac{1}{2}$ ounce of alcohol and $\frac{1}{2}$ ounce of water. Either of these liquids are applied by rubbing into the skin. The application should be repeated every 4 or 5 days until the disease is cured.

Other Mites Affecting Poultry

Another form of *Body Mange* or scabies is found associated with the mites *Epidermoptes bilobatus* and *Epidermoptes bifurcatus*, but it has not been certainly demonstrated that they are the cause of the disease. Present evidence indicates that they are.

The disease closely resembles *favus* (p. 223), but usually does not affect the head. The regions commonly attacked are the neck, breast, the wings and the body under the wings. It sometimes affects the entire body, including the head. The skin becomes irritated and shows an accumulation of scales or crusts especially at the base of the feathers.

The mites live on the skin at the base of the feathers. Since the mites are sometimes found on birds which show no signs of scabies and since the disease so closely resembles favus, which is known to be caused by a fungus, it is sometimes supposed that this mange is also due to a fungus and that the mites are inoffensive.

Five species of mites have been recorded which live upon the feathers of fowls. These are fairly abundant but do no harm.

Two mites live within the body of fowls. One of these, the air-sac mite, is described elsewhere (p. 180). The other, the connective tissue mite, *Symplectoptes cysticola*, is found in the connective tissue of the fowls. They produce local irritations giving rise to tubercles, but apparently do not affect the health of the bird.

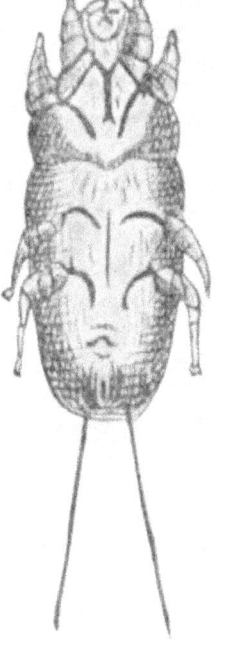

Fig. 47. — *Symplectoptes cysticola.* Connective tissue mite. (After Theobald.)

The larvæ of the so-called "harvest-bug" (which is not a *bug* at all), *Tetranychus* (*Thrombidium*) (*Leptus*) *autumnalis*, sometimes attacks poultry. The appearance of this mite is shown in Fig. 48.

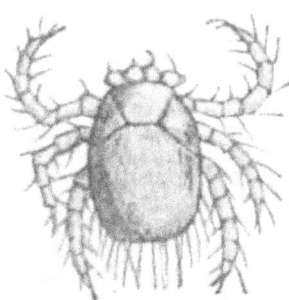

Fig. 48. — "Harvest bug," *Tetranychus (Leptus) autumnalis,* larval form. (After Murray.)

This small brick red mite, barely visible to the naked eye, is bred upon berry and currant bushes, vegetables and grain, but when opportunity offers it bites almost any animal, often attacking man. It sometimes causes considerable mortality among late hatched chickens which frequent its breeding places. The parasites fasten themselves so firmly by their claws and palpi that they can only be detached by force. They produce intense irritation, which

228 *Diseases of Poultry*

often leads to epileptiform symptoms and death follows in a few days.

Theobald suggests dusting flowers of sulphur among the feathers when the parasites are present. Probably the Lawry lice powder (p. 211) would be more effective. When

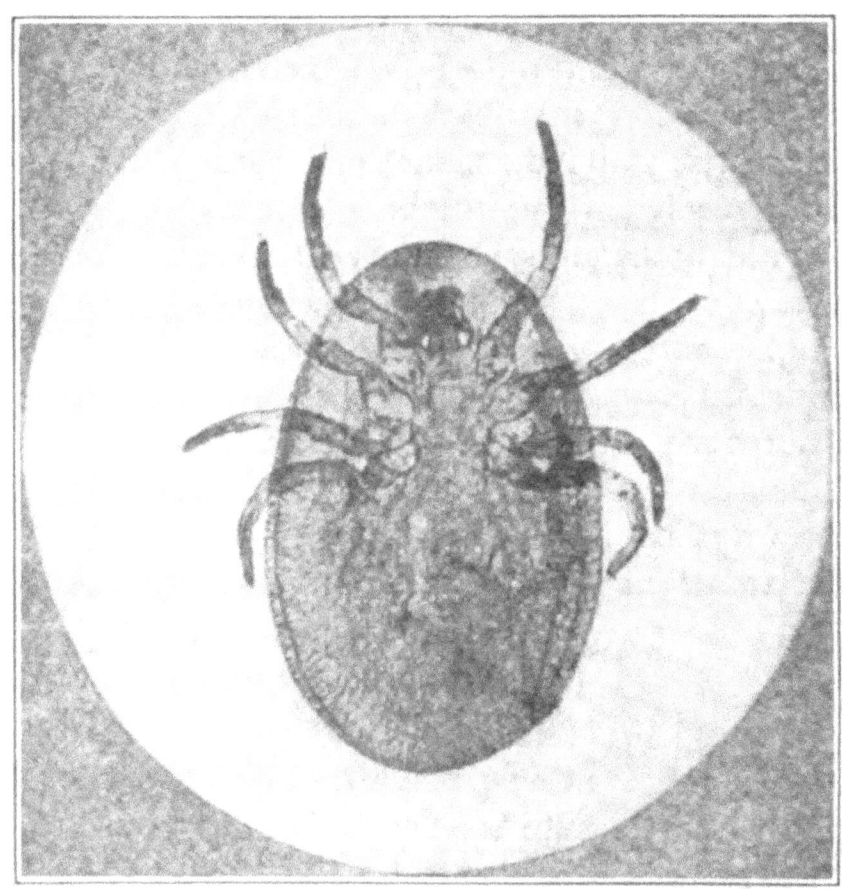

Fig. 49. — The poultry tick, *Argas persicus*. Adult ventral view, showing the four pairs of legs, mouth parts, etc. (After Laurie.)

these parasites are abundant chickens should be kept away from the places where the mites breed.

Ticks. — A poultry tick, *Argas persicus*, occurs in South Africa, Australia, and many other parts of the world. It occurs to some extent in the southern part of the United States. Where present it is an exceedingly destructive pest. The following notes are taken from a paper by Laurie.[1]

These parasites belong to the group of mites *Acari*. In the adult stage they have four pairs of legs but in the larval stages only three pairs (Figs. 49 and 50).

These ticks are nocturnal in habit. During the day they secrete themselves in cracks and crevices and are rarely seen. At night they come out on to the roosts and fasten

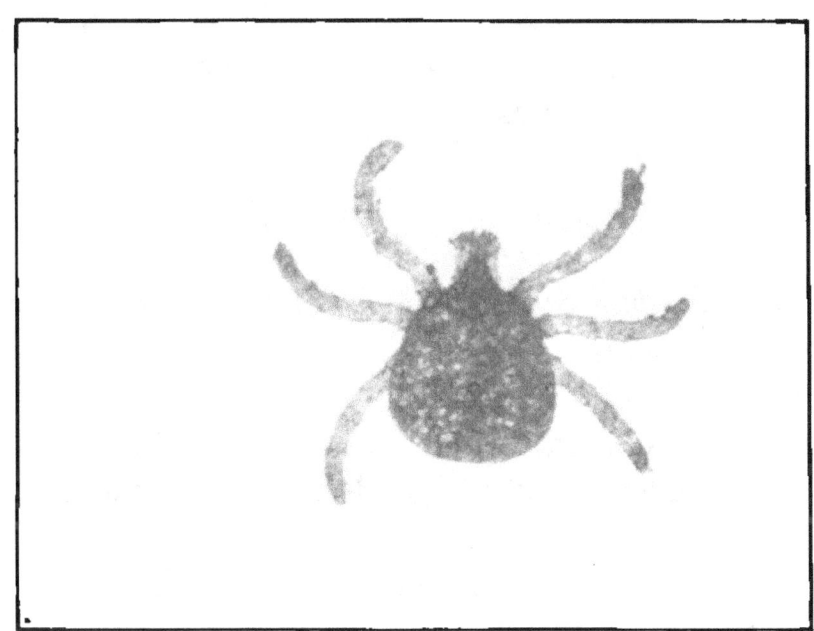

FIG. 50. — The poultry tick. Larva, showing the three pairs of legs. (After Laurie.)

themselves upon the birds. After gorging themselves with blood they return to the cracks to digest their meal. An adult tick feeds only about once a month, as it requires that time to digest fully one meal. During the growing periods they undergo a molt after digesting each meal. These ticks breed very prolifically, so that a poultry house once infested soon becomes overrun by them.

[1] Laurie, D. F. The Poultry Tick. Dept. of Agric. of South Australia, Bul. No. 74, pp. 1–32, 1912.

A considerable portion of the injury done by these parasites is due to the irritating annoyance of the feeding ticks. It has been found, however, that this is not the most serious injury. In those countries infested with these ticks there is a disease known as the tick fever. It has been shown that this disease is caused by a protozoön blood parasite, *Spirochæta marchouxi*, and that this protozoön lives in the poultry tick as an intermediate host.

Treatment.—Thorough cleanliness and disinfection are the remedies to use against this tick. Five to ten per cent kerosene emulsion applied to the clean roosts and walls will rid the place of these pests.

Other External Parasites

The *dove cot bug* or "bed-bug" of poultrymen, found in pigeon lofts, sometimes invades neighboring hen roosts. It probably sometimes attacks fowls. It resembles closely the bed bug found in dwelling houses, and like this pest is hard to exterminate as it can live almost indefinitely on dead organic matter. This tick hides in cracks during the day and attacks its host only at night. Persistent repetition of the sprays recommended for hen roosts infected with red mites (p. 216) will destroy these parasites.

Leaflet No. 57 of the English Board of Agriculture gives the following brief account of the hen flea, *Pulex gallinæ* (or *avium*):

"The fleas, which are true insects, belong to the order of flies (*Diptera*). They feed upon the blood. One species only lives upon the fowl, namely the bird flea (*Pulex gallinæ* or *avium*) which attacks also most other birds. The hen flea, as it is generally called, is abundant in dirty fowl runs, and especially in the nests where straw is used. The adult flea is dark in color, and, as in all fleas, is devoid of wings.

The fleas are provided with very sharp piercing mouths. They are what are termed 'partial parasites'—parasites that only go to their hosts to feed. The fleas are not noticed on the birds because they generally attack them at night; then, however, they do much harm, causing constant irritation and loss of blood, and depriving them of rest.

"*Life-history of Hen Flea.* — The female flea lays her eggs (nits) chiefly in the nests amongst dust and dirt and in the crevices of the walls and floor. These nits give rise to pearly white maggots, with brown horny heads, which can often be found in the bottom of the nests amongst the dust. These larvæ are mature in 2 or 3 weeks, then they reach about ⅛ of an inch in length. In warm weather they may be full fed in even 10 days.

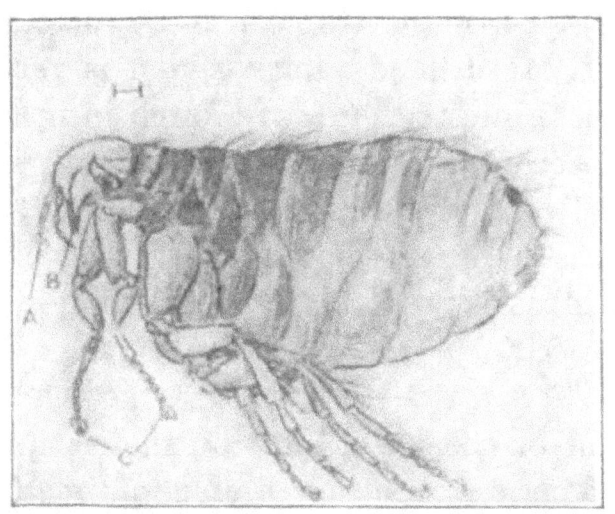

FIG. 51. — The chicken flea, *Pulex gallinæ* or *avium*. The mark above the head indicates the actual size. (After Kaupp.)

They then spin a pale cocoon amongst the dirt, in which they pupate. The pupa is at first pale brown, then dark chestnut brown. In this condition the flea remains 10 to 21 days, when the pupa hatches into the adult. They breed all the year round, but chiefly in warm weather. It is well to remember that, whenever there are dark and dirty hen roosts, there are sure to be a number of *Pulex gallinæ*."

Treatment. — These parasites do not usually occur under sanitary housing conditions. When they occur the houses should be cleaned and sprayed as for red mites (p. 215).

Theobald recommends the use of excelsior or shavings instead of straw for nesting material, as the fleas do not breed as readily in this material.

Manson's Eye Worm

This parasite (*Oxyspirura mansoni*) was first reported from America in 1904.[1] It appears to occur very infrequently in this country at the present time. It is abundant in some of the tropical and subtropical countries. Wilcox and McClelland[2] state that it is very common in Honolulu where infested birds are found in nearly every flock.

According to these writers the eggs are laid in the eye and are washed down the lacrymal duct and thence to the intestines. The eggs hatch and the larvæ live until half grown on damp soil. At this time they enter the eye of the chicken directly.

Treatment. — Anesthetize the eye with 5 per cent solution of cocaine. Lift the nictitating membrane and drop a 5 per cent solution of creolin directly into the inner corner of the eye under the membrane.

Liming the soil in the yards or keeping the birds on dry, frequently disinfected floors until the infestation disappears are recommended.

[1] Ransom, B. H., "Manson's Eye Worm of Chickens," etc. U. S. Dept. of Agr., Bur. of Anim. Ind., Bul. 60, pp. 72, 1904.

[2] Wilcox, E. V., and McClelland, C. K., "The Eye Worm of Chickens." Hawaii Exper. Sta. Press Bul. 43, p. 14.

CHAPTER XVI

Diseases of the Skin

Favus (Baldness or White Comb)

This disease of the skin attacks poultry as well as man and the domestic mammals. In mammals it is called *tinea favosa* or *favus*.

Diagnosis. — The disease usually appears first as small gray white spots on the comb, wattles, eyelids and around the ears, that is, on the unfeathered parts of the head. The spots enlarge and run together forming a scaly crust which becomes thicker until in three or four weeks it may be as much as 8 millimeters ($\frac{1}{3}$ inch) thick. The scales which make up the crust are often formed in concentric rings, the margins raised and the centers depressed, so that the scale is somewhat cup shaped. When the crust is removed the skin appears irritated and in places the surface is somewhat raw.

Fig. 52. — Head and neck of a fowl affected with generalized favus. (After Pearson.)

The disease spreads to the feathered parts of the head, the neck and the region around the vent. The base of the feathers becomes surrounded by concentric rings of the scaly material. The feathers become dry, erect, and brittle and finally break off or fall out leaving a disk-shaped scale with a depression at the bottom where the base of the feather was located. The bird's head and neck and patches around the vent become bare of feathers. The exposed skin is covered with the cup-shaped scales. Sometimes the disease spreads over the whole body until the bird becomes nearly naked. The diseased bird has a peculiar disagreeable odor, sometimes likened to the odor of a musty grain or to moldy cheese and sometimes to cat's urine or to macerating animal material. In early stages the general health does not appear to be affected, but as the disease advances the bird loses its appetite, becomes poor and exhausted, and finally dies.

Etiology. — The disease is caused by the fungus *Achorion schonleinii.*

This fungus is found in the cup-like scales on the skin and in the quills of the feathers of the diseased parts. If the favic cups or scales are moistened with weak acetic acid and examined under the microscope, it will be seen that they are formed of branching, thread-like mycelial tubes of the fungus closely interwoven with one another, spores of the fungus, and epithelial scales from the skin of the host embedded in a viscid substance secreted by the fungus. Some of the tubes of the mycelium contain spores. Many of the spores are found free among the filaments. They are usually found in groups of 3, 4, or 8.

Both the mycelium and spores of the fungus are found in the quills of the feathers of the diseased parts. The fungus sometimes penetrates even the barbs of the feathers.

Favus is a contagious disease and gets into a flock by the introduction of an affected bird. It is less likely to attack strong, vigorous birds than those in poor condition. It usually starts at a point where the skin is broken. Young birds are more susceptible than old ones. The large Asiatic breeds are specially liable to take the disease. No breed is entirely immune.

Mégnin [1] and some other authors consider this disease distinct from the favus of man and other animals, but numerous recorded observations indicate that it is the same disease, and may be communicated to man. In handling affected birds, therefore, care should be exercised to prevent infection of cuts or scratches.

Treatment. — Diseased birds should not be introduced into a flock. If the disease has been accidentally introduced the affected birds should be isolated as soon as possible. The flocks should be watched in order to discover and isolate any new cases that appear.

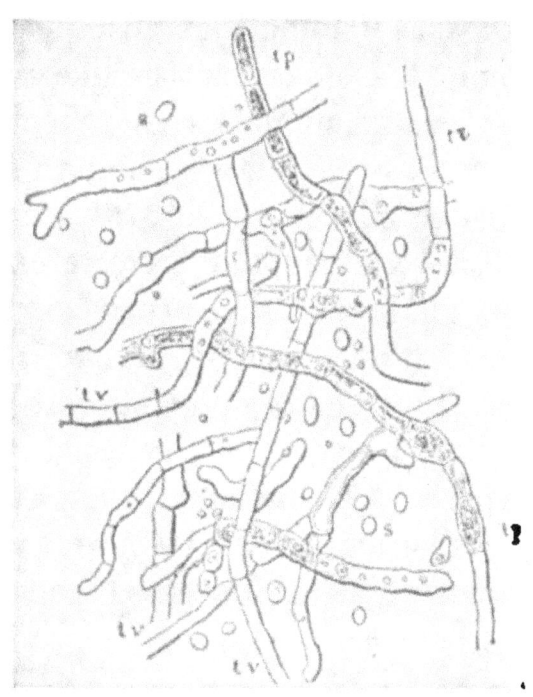

FIG. 53. — The fungus *Achorion schonleinii* which causes favus in poultry. *tv.* Empty tubes of mycelium. *tp.* Tubes of the mycelium containing protoplasm and spores.

In early stages the disease yields readily to treatment. Zürn considers treatment economically advisable only before the feathered parts of the body are attacked. The disease may sometimes be cured at a later stage. The value of the affected bird must determine whether or not it is worth treating.

As much of the crust as possible should be removed. This

---

[1] Mégnin, P., "Médecine des oiseaux." Vincennes, 1906.

is best done by first softening the scabs with warm water or with oil or glycerine. Robinson recommends scraping with the back of a knife or a spoon handle. The parts should then be painted with tincture of iodine or should be bathed with corrosive sublimate solution, 1 part of the sublimate to 1000 parts of water, and then rubbing with the ointment described on page 55. In using the corrosive sublimate solution it should be borne in mind that this solution, which unless colored with some dye looks exactly like water, is extremely poisonous to men and animals when taken internally. Dishes or bottles of corrosive sublimate should never be left where they can be accidentally mistaken for water.

Lard and sulphur are often used successfully in the treatment of favus. Use nearly as much sulphur as lard and work them into a smooth salve. In early stages the disease usually yields to application of lard or oil alone.

Prognosis. — In early stages the favus may be cured at the expense of a small amount of attention. After the feathered parts become affected a cure requires considerable labor as the fungus is better protected from the applications.

White Comb

This name is often used for favus, but some authorities (*e.g.*, Vale) use it to designate a condition of the comb characterized by a white powdery scurf of the surface. The comb is light colored and the white scales or flakes are particles detached from the epidermis. This condition is thought to be due to anæmia. Wright says that it "appears generally due to dirt, or overcrowding in small space, or want of green food." The only treatment advised is to place the birds under sanitary conditions and give them a good balanced ration.

Chicken Pox (Sore-head or Epithelioma Contagiosum)

This contagious disease of poultry, although widely distributed in the northern states, is less common and serious here than in the Gulf states and Hawaiian Islands. It is impossible at present to decide whether this is a distinct disease or a form of roup which affects the skin of the head. This can only be determined when further investigations have revealed the real cause of these diseases. Experiments have shown that both pox and diphtheria are easily inoculated from fowl to fowl and from pigeon to fowl, but the inoculation of pox from fowl to pigeon has proved very difficult and that of diphtheria impossible. The contagion is believed to exist in the blood as well as in the nodules which appear upon the skin.

The disease is generally introduced by new birds which are put into the flock or by exhibition birds which return infected. Probably it is often brought by pigeons, sparrows, and other birds which fly from one yard to another. The inoculation of the comb and wattles appears to occur by rubbing these parts with the infected feet or by being injured with the infected beaks of other birds.

The virus is quite resistant and requires thorough disinfection for its eradication.

Diagnosis. — The disease usually appears as warty nodules on the unfeathered parts of the head. They look like the tumors in the nasal passages and eye sockets of birds affected with roup.

Freidberger and Frohner [1] give a good description of these nodules on the skin of the head, as follows:

"Their favorite seats are those parts of the head that are not covered with feathers; root of the beak, neighborhood of

---

[1] Freidberger and Frohner, "Veterinary Pathology." (Vol. I. Hayes transl.) Quoted from Cary.

the nostrils, angles of the mouth, lobes of the ear, parts adjacent to the auditory meatus, wattles, surface of the face, edges of the eyelids, intermaxillary space, and especially the comb. They sometimes spread over the feathered parts of the head, throat and neck, and may occur on the outer surface of the thighs, abdomen, under the wings and in the vicinity of the cloaca. At first these epitheliomata appear in the skin, as flat nodules, which soon become prominent, and which vary in size from a poppy seed to a millet seed. Later on, they usually attain the size of a hemp seed. They are of a reddish-gray or yellowish-gray color, often show distinctly in their earlier stages of development a peculiar greasy, nacreous luster; and are rather firm to the touch. Their surface soon becomes covered with a dirty-gray, yellowish-brown or red-brown crust. They are discrete and disseminated in considerable numbers on the erectile tissues, etc. They vary in size according to their age; and frequently lie rather close to one another, so that the affected parts look as if coarsely granulated; or they are crowded together in such a manner as to give the appearance of large warts with divisions through them, or mulberry-like hypertrophies. Even single nodules, to say nothing of the groups, may attain the size of a lentil, pea, cherry-stone, broad bean or larger object. The older they become, the rougher and more covered with knobs will be their incrusted surface.

"If the edges of the eye-lids be affected by these tumors, the lids will become nodular, swollen and closed. The conjunctiva in this case also suffers; it projects outwards because catarrhally inflamed; assumes a yellowish color at the seat of eruption; and its surface becomes covered with crusts. Purulent conjunctivitis may appear and the inflammation may spread to the sclerotic and cornea, with keratitis and panophthalmia as the result. If, as sometimes happens with pigeons, the eruption of nodules extends over the whole of the skin of the eye-lids and its neighborhood, the entire eye will become covered with mulberry-like proliferations of various sizes."

The presence of these nodules on the epithelium of the head is often (but apparently not always) accompanied with

Fig. 54. Sore-head on comb, eyelids and skin. (After Hadley and Beach.)

characteristic roup lesions of the nasal cavities, mouth and throat. As long as the disease is confined to the skin of the head the general health of the bird does not seem to be affected. Recovery may take place without treatment in from 10 to 20 days. The nodules in such cases dry up and fall off. Usually, however, the disease is not self-limited, but advances. The eyes may become closed so that the birds cannot see to eat. They get poor and die from exhaustion. When the mucous membrane of the mouth develops diphtheritic membranes, death occurs earlier than in other forms.

Etiology. — The lesions of this disease resemble the lesions of roup and many of the same micro-organisms are found in the two cases. The organisms isolated from the lesions of sore-head include several bacteria, a coccidium, a yeast and several molds. The coccidium, one of the molds, and one of the bacteria have each been considered the cause of the disease by different workers. The real cause of the disease and its relation to roup must be determined by further investigations.

Many recent investigations have indicated that the disease is caused by a filterable virus.[1] In regard to the etiology and mode of transmission Cary[2] says:

"It is evidently infectious; because the disease in all its forms, spreads rather rapidly from one chicken or pigeon to another. Ward, Harrison and others have transmitted, in some cases quite readily by carrying a small amount of diseased material (exudate and blood), from a sore-head chicken to healthy chickens. It is also quite certain that chicken pox and pigeon pox are identical or one and the same disease.

"Mosquitoes, gnat flies, chicken mites (ticks), chicken lice, chicken foot mites (*Sarcoptes mutans*) and possibly cockroaches may sometimes be carriers of the real virus. It seems quite certain that mosquitoes can transmit the virus from

---

[1] For example see:

Bruet, E., "Contribution a l'étude l'epithélioma contagieux des oiseaux." *Ann. l'Inst. Past.*, T. 29, pp. 742–765, 1906.

Sweet, C. D., "A Study of Epithelioma Contagiosum of the Common Fowl." *Univ. of Calif. Public Zool.*, Vol. II, pp. 29–51, 1913.

[2] Cary, C. A., "Chicken-pox or Sore-head in Poultry." **Alabama** Agr. Expt. Stat. Bul. 136, pp. 17–56, 1906.

water or some other source, under certain conditions. Warm and wet weather seem to increase the virulency of the virus and favor the rapid transmission of the disease. It is not impossible that ants may have a rôle to play in the transmission or cause of sore-head.

"*Pathological Anatomy.* — On the skin the small, greasy-like nodules, or hypertrophied nodules of the skin, contain epithelial cells that have in them 'greasy' refractive bodies that stain yellow with picro-carmine and the nuclei of the epithelial cells become 'reddish brown' in color. Nearly all the epithelial cells in the nodule appear larger than normal and contain the refractive bodies. In the younger epithelial cells these bodies (young coccidia) are relatively small and occupy $\frac{1}{4}$ to $\frac{1}{3}$ of the epithelial cavity. In the older or outer or cast-off epithelial cells these refractive bodies are said by Freidberger and Frohner to occupy the entire cavities of the epithelial cells. The invaded or infested epithelial cells are unusually larger than the epidermal cells of the healthy neighboring skin. Among the cast-off mass of epithelial cells are found round refractive bodies and numerous nuclei of leucocytes or pus cells. The subcutaneous connective tissue is hyperæmic (congested) and is infiltrated with cells (leucocytes and nuclei of disintegrated cells). Possibly some of the small nuclei-like bodies among the cells in the subcutis may represent one stage in the development of coccidia. Many observers have, also, found various bacteria in the nodule and subcutis.

"When the exudate on the mucous surface or the crust of the nodule of the skin is torn off the raw surface bleeds rather freely and a fresh mount of this blood contains a short oval bacillus, numerous round bodies usually said to be nuclei of leucocytes; and a few polynuclear leucocytes. Repeated inoculations in the comb, wattles, skin and conjunctiva and oral mucosa of healthy chickens of various ages, with this blood, fresh from under a nodule or a diphtheritic exudate, has failed to produce positive infective results. I have also tested it on pigeons with like negative results.

"The exudates on the mucous membrane of the throat, mouth or larynx appear to be very much alike in all forms of the disease."

The *period of incubation* is said to vary all the way from 2 to 20 days. Cary reports a case in which a newly purchased Barred Plymouth Rock cockerel was placed in the pen with chickens just recovering from the disease. In 24 hours this bird had developed a well marked case of sorehead on the wattles, comb, and eyelids. Apparently the period of incubation varies with the mode of transmission, virulency of the virus, the weather (rapid in damp, warm weather and slower in cool and dry weather), and the age and condition of the chicken or pigeon. Chicks from broiling size up to 7 or 8 months old seem to be most susceptible. Chickens with large combs seem to be more susceptible than birds with small combs and wattles.

Treatment. — The introduction of diseased birds into healthy flocks should be avoided. The same precautions should be practiced in the isolation of sick birds and disinfecting the houses as is advised for roup (p. 161). When the disease is localized a small amount of individual treatment cures many cases. The crust or nodules should be removed and the places treated with creolin (2 per cent solution) or corrosive sublimate ($\frac{1}{1000}$) (p. 54) and dusted with iodoform. The iodoform may be put into the eye. When the disease is not far advanced one such treatment may be followed by daily greasing with the ointment recommended on p. 55 or with vaseline or lard. In bad cases the iodoform should be used daily for a few days and then the ointment.

Recently some work has been done on the production of

artificial immunity to this disease. Manteufel[1] used a virus obtained by scraping off the softened epitheliomæ and macerating these in physiological salt solution. Chickens injected intravenously or subcutaneously with this virus showed an immunity which lasted from one and a half to two years. He also tried a hyperimmune serum but without success. Hadley and Beach[2] report good success with a vaccine prepared in the same way. Their vaccine was prepared by grinding pock scabs, diphtheritic membranes, etc., in a sterile mortar with physiological salt solution. This was then filtered through cotton and heated for one hour at 55° C. In an infected flock 440 healthy birds were vaccinated with two doses of 1 cc., each at an interval of five days. Only four birds, or less than 1 per cent, in this lot developed noticeable symptoms. Of 75 control birds, 26 cases developed in three weeks.

Hadley and Beach recommend this vaccine treatment as "especially applicable in large commercial and experimental flocks, where the greatest losses are sustained and where preventative measures can be most economically carried out. Breeders of pure bred and fancy fowls whose stock would be impossible to replace should also find it valuable."

Prognosis. — The mortality among birds affected with this disease is said to be from 50 to 70 per cent. Cary (*loc. cit*) says: "I judge this a low per cent of losses if birds are left to themselves without proper care or treatment. But if individual treatment is patiently and regularly applied the mortality can be cut down to less than 20 per cent. If only the skin of the head, and the comb and wattles are involved,

---

[1] Manteufel, *Arb. Kaiserl. Gesundheitsamt.* Bd. 33, pp. 305–312, 1910.

[2] Hadley, F. B., and Beach, B. A., "Controlling Chickenpox, Sorehead or Contagious Epithelioma by Vaccination." Proc. Amer. Vet. Med. Assoc., Vol. 50, pp. 704–712, 1913.

one should lose less than 10 per cent. . . . But if the nasal passages and trachea are involved, or the intestines become involved, — good care and treatment may save 50 to 80 per cent."

Edema of the Wattles

Seddon[1] has recently described a disease of the wattles of fowls showing two very marked symptoms, namely (1) enlargement due to the presence of the inflammatory fluid and later (2) distortion with the formation of hard nodules of cheesy material in the wattle. He believes the disease is a localized form of fowl cholera, in which the causative organism gains entrance to the wattles through scratches, etc. Septicæmia and death occur in a certain percentage of the cases. The disease usually runs a chronic course with subsequent replacing of the edematous fluid by fibrous tissue and results in the wattle assuming a crinkled appearance.

Treatment. — "Cropping" of the wattles is recommended in some cases. The adoption of general sanitary measures is of most service in suppressing the disease. (See Chapter II.)

[1] Seddon, H. R., "A Disease of the Wattles of Fowls." *Jour. Dept. Agr. Victoria*, Vol. 12, pp. 426–428, 1914.
———, "Edema of the Wattles of Fowls Due to an Organism of the Pasteurella Group." *Vet. Jour.*, Vol. 70, pp. 24–34, 1914.

CHAPTER XVII

Diseases of the Reproductive Organs

The direct economic importance of poultry lies in the production of two things, viz., meat and eggs. For the production of the latter the poultryman is dependent upon the activity of the reproductive system of the hen. Under natural conditions in the wild state, the progenitors of the domestic fowl laid relatively few eggs. Judging by other species of wild birds of the present day, however, it is highly probable that the wild progenitors of poultry possessed the potential ability to lay much more than the usual number of eggs provided they were removed from the nest as fast as laid. Under domestication this practice of removing the eggs as fast as laid, together with the feeding of rich foods, and still other factors, lays heavy demands upon the reproductive system. It is not remarkable that an organ system which under conditions of nature produced from 12 to perhaps 30 units per annum, frequently breaks down under the strain of producing from 100 to 250 per annum of the same kind of units. It could only be expected that, as is actually the case, the egg producing organs would be particularly liable to disease.

ANATOMY AND PHYSIOLOGY

In order that the discussion of the diseases of the reproductive organs may be intelligible it is desirable to preface it with a brief account of the anatomy and physiology of the organs of reproduction in the hen. Because of the fact that the corresponding organs in the male are less subject to disease, on the one hand, and are perhaps better understood by the poultryman, because of the prevalence of the practice of caponizing, on the other hand, it will not be necessary to discuss the male in detail in this connection.

The organs concerned in egg production in the hen are shown graphically in Fig. 55. This picture and the accompanying explanation of it will make clear the various parts of this organ system. All of the points shown in the figure may easily be demonstrated on a hen, killed during a period of laying activity. It should be noted that this picture is somewhat diagrammatic and not in accord with normal conditions in respect to at least two points. These are: (1) there are two eggs in the upper portion of the oviduct. Normally there would be but one there at a time. (2) The proportionate lengths of albumen portion, isthmus and uterus are not correctly indicated.

In this figure the various numerals have the following significance:

1. The *ovary*; region in which the ovules (later to become yolks) are still small in size.

2. An *ovule* in an intermediate stage of development, larger than those at 1, but still not ready to pass into the oviduct to be laid. It is contained in a very vascular capsule, known technically as the *follicle*.

3, 3. Ovules still larger and containing more yolk. The lower one is nearly ready to leave the ovary and pass down the oviduct.

4. It will be noted that on all the larger follicles there is one region (forming a line) in which there are no blood vessels. This region (4, 4) is known as the *stigma*. Here the follicle wall breaks and allows the ovule (yolk) to leave the ovary preparatory to laying.

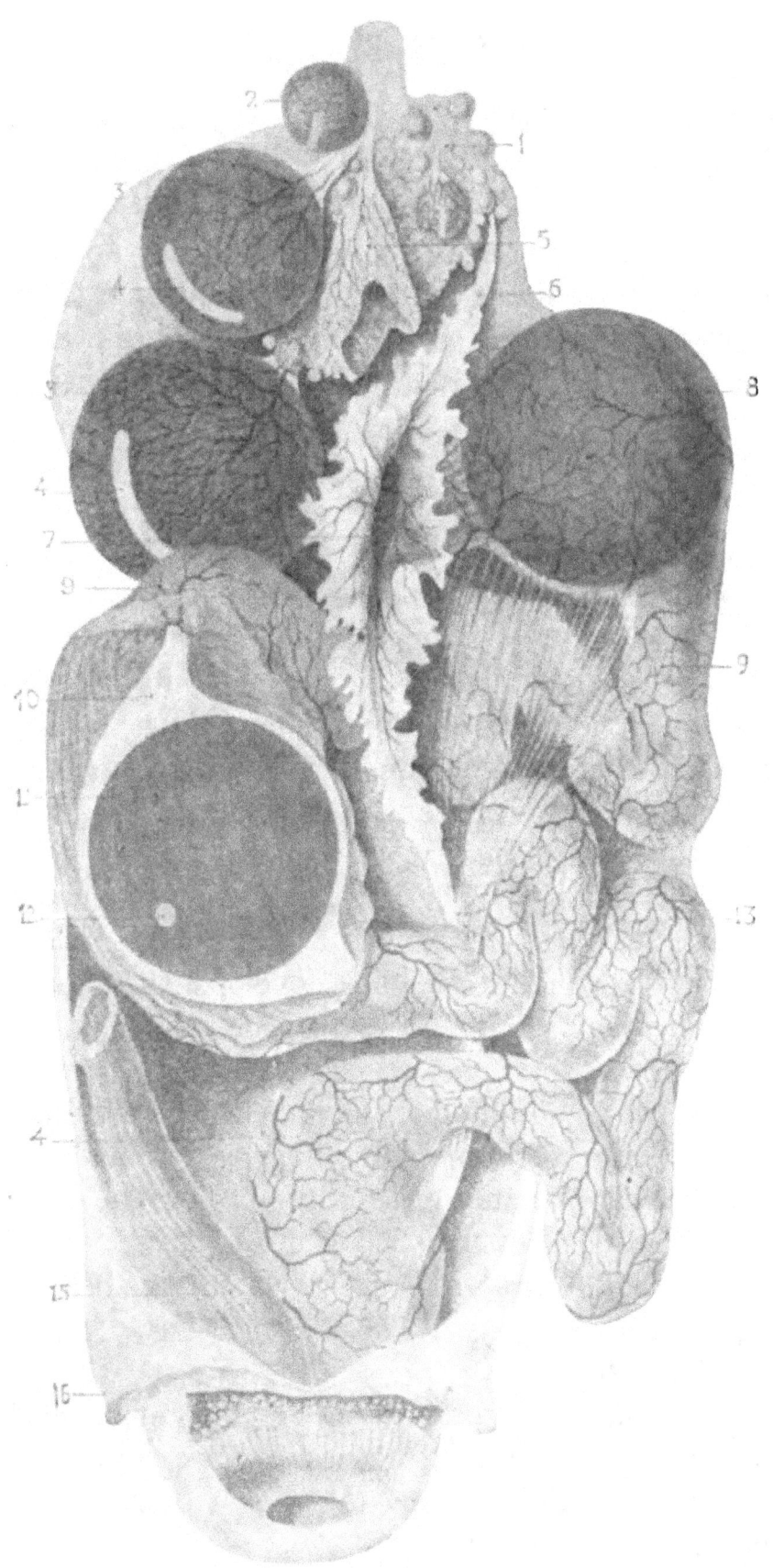

Fig. 55. — The reproductive or egg producing organs of a hen. See text for explanation of figures. (After Duval.)

5. An empty *follicle* in which the *stigma* has opened and the yolk passed out.

6. Anterior end of the margin of the *funnel* (or *infundibulum*) of the *oviduct* or egg-tube. When an ovule is about to be discharged from the ovary these funnel lips or margins wrap around that portion of the ovary, so that the ovule may certainly pass into the oviduct and not into the abdominal cavity.

7. Opening of the *funnel*. Through this opening the yolk passes into the oviduct.

8. A yolk which has just passed through the funnel opening into the upper portion of the oviduct.

9, 9. *Albumen secreting portion* of the oviduct in which the greater portion of the albumen or white of the egg is secreted by glands in the walls of the oviduct in this region.

10. First layer of albumen, or white, secreted about the yolk. From this layer are formed the *chalazæ*, or cords of twisted, thickened albumen, at each pole of the yolk.

11. Yolk, around which albumen is being secreted.

12. The *germinal disk*. This is the living portion of the egg, from which the future chick develops, the main mass of yolk serving as food material for the developing embryo during the process.

13. Anterior end of the *isthmus* of the oviduct. The primary function of the isthmus is to secrete about the egg the *shell membrane*, the dense white membrane closely adherent to the inside of the shell of an egg.

14. The *uterus*, or *shell gland*, in which the shell is put on the egg.

15. The *rectum*.

16. The walls of the abdomen cut and folded back.

17. External opening of the *cloaca*, or common space into which open (*a*) the rectum, (*b*) the oviduct and (*e*) the ureters, or kidney ducts.

The processes concerned in the formation of an egg may be summarized as follows:

Certain ones of the small oöcytes in the ovary (Fig. 55, 1) are all the time coming into a state of physiological activity, while the hen is in a laying cycle. These oöcytes grow in size by the deposition of yolk until finally they are of the full size for laying. The time required for this final growth of yolks preparatory to laying is not far from 20 days, on the average.[1]

The fully formed yolk, or *ovum*, leaves the follicle through a rupture of the latter along the *stigma* (Fig. 55). This process is called *ovulation*. As it leaves the ovary the ovum is received by the funnel of the oviduct.

After entering the infundibulum the yolk remains in the so-called albumen portion of the oviduct about three hours and in this time acquires only about 40 to 50 per cent by weight of its total albumen and not all of it as has hitherto been supposed. During its sojourn in the albumen portion of the duct the egg acquires its chalazæ and chalaziferous layer, the dense albumen layer, and (if such a layer exists as a distinct entity, about which there is some doubt) the inner fluid layer of albumen.

Upon entering the isthmus, in passing through which portion of the duct something under an hour's time is occupied instead of three hours as has been usually maintained, the egg receives its shell membranes by a process of discrete deposition. At the same time, and during the sojourn of the egg in the uterus, it receives its outer layer of fluid or thin albumen which is by weight 50 to 60 per cent of the total albumen. This thin albumen is taken in by osmosis through the shell membranes already formed.

---

[1] Cf. Gerhartz, W., "Über die zum Aufbau der Eizelle notwendige Energie (Transformationsenergie)." *Pflüger's Arch.*, Bd. 156, pp. 1–224, 1914.

When it enters the egg in this way it is much more fluid than the thin albumen of the laid egg. The fluid albumen added in this way dissolves some of the denser albumen already present, and so brings about the dilution of the latter in some degree. At the same time, by this process of diffusion, the fluid layer is rendered more dense, coming finally to the consistency of the thin layer of the laid egg. The thin albumen *layer*, however, does not owe its existence in any sense to this dilution factor, but to a definite secretion of a thin albumen by the glands of the isthmus and uterus.

The addition of albumen to the egg is completed only after it has been in the uterus from five to seven hours. Before the acquisition of albumen by the egg is completed a fairly considerable amount of shell substance has been deposited on the shell membranes. For the completion of the shell and the laying of the egg from twelve to sixteen, or exceptionally even more, hours are required.[1]

The main factor in propulsion of the ovum along the oviduct appears to be the peristaltic movements of the latter; it is probable that the cilia which line the cavity have something to do with the rotation of the ovum on its chalazal axis.

With this account of the anatomy and physiology of the female organs of reproduction in hand we may proceed to a consideration of their diseases. These diseases fall at once into two classes: (*a*) those affecting the ovary, and (*b*) those affecting the oviduct.

[1] The foregoing account is based upon that given by Pearl and Curtis, *Jour. Exper. Zool.*, Vol. 12, pp. 123 and 124, 1912.

DISEASES OF THE OVARY

Atrophy of the Ovary

By "atrophy" of the ovary is meant a diminution in size of that organ accompanied with a cessation of its physiological activity. It may shrink to the size and appearance which it has in a very young bird. The following sorts of atrophy of the ovary may be distinguished. The different sorts are separated from each other, not because of any difference in the end result, but because of the different etiological factors concerned.

1. Physiological atrophy.
    a. Temporary.
    b. Permanent.
2. Congenital atrophy (Pseudo-hermaphroditism).
3. "Black atrophy."

A physiological diminution in size or partial atrophy of the ovary occurs normally in fowls when after a period of laying they go into a more or less prolonged resting period. This condition of the ovary is usually (in fowls under two years old) only temporary. The organ resumes its normal size and activity after a time. In old birds (3 to 6 or more years of age) it not infrequently happens that the ovary passes into an atrophied condition, and remains permanently in that condition thereafter. In such cases the bird as a whole, and the ovary in particular, may be perfectly healthy, showing no sign of disease. Cases of permanent physiological atrophy of the ovary have been observed by the writers as follows:

One case in a White Crested Black Polish.

One case in a Cornish Indian Game.

Several cases in Barred Plymouth Rocks. All of the latter were birds of very high fecundity (200 or more eggs per annum) in their pullet years.

It should be noted that in what is here called permanent physiology atrophy of the ovary there is *no* associated change of the secondary sexual characters. That is, the hen does not assume cock plumage, spurs, enlarged comb and wattles, nor any other of the secondary sexual characters normal to

Fig. 56. — Showing a case of incomplete hermaphroditism. In front of the line *ab* the bird has the characters of the male, behind it the characters of the female. The ovary was not functional in this bird. (Original.)

the male. This indicates that in permanent physiological atrophy (just as is known to be the case in temporary) the only function of the ovary which is disturbed is that which is involved in egg formation. The activity of the organ in regard to producing an internal secretion which in some way controls the secondary sexual characters remains unchanged.

As *congenital atrophy* of the *ovary* are to be classed cases of psuedo-hermaphroditism in fowls. In such cases a true, functioning ovary never develops. There may be a body which in gross features resembles an ovary, but it is inactive and does not take even the first steps in oögenesis (egg formation).

There may or may not be a testis like body present in these cases. Not only is the egg producing activity absent in such cases, but also in many of them at least, the internal secretion normally produced by the ovary is lacking also. The bird then takes on some or all of the secondary sexual characters of the male. The appearance of such a bird is shown in Fig. 56.

As "black atrophy" of the ovary is here designated the peculiar diseases of the ovary first observed more than a century ago in England as occurring in pheasants. The striking feature of the disease is that under its influence the bird assumes the plumage appropriate to the male. The change in the ovary and oviduct induced by the disease appears to be an atrophy accompanied by a blackening which is probably a true melanosis. The following account of an outbreak of this disease about fifty years ago by Hamilton [1] is of interest. "In the years 1858, 1859, and 1860 this peculiar alteration of structure in the female organs of generation in the Pheasants was particularly prevalent in some parts of England. I had the opportunity of examining many specimens, and was able completely to confirm Mr. Yarrell's views on this subject. Indeed, the majority of the birds were young females, many of them being birds of the year, some being in their first molt. I found also that the plumage varied and approached that of the male, not in accordance with the age of the bird, but with the

[1] Hamilton, E., "On the Assumption of the Male Plumage by the Female of the Common Pheasant." Proc. Zool. Soc., London, 1862, pp. 23–25.

amount of disease of the generative organs. The greater the destruction of the ovarium and oviduct, the nearer the plumage assimilated that of the male.

"For example, in birds with the hen-plumage predominating, the ovarium and oviduct exist as in the fecundating hen, the small ova lying in considerable numbers in the ovarium, the ovarium and oviduct showing dark lead-colored masses of disease.

"In birds with the plumage of the male in a measure exceeding that of the female, the ovarium is considerably diminished in size, dark-colored, and containing only a few blackened ova; the oviduct is spotted with dark patches, and considerably contracted.

"And thirdly, in birds with the male plumage predominating over that of the female, the ovarium is reduced to a small dark amorphous mass, resembling the coagulated blood, the presence of ova cannot be detected, and the oviduct is almost entirely obliterated at its junction with the ovarium. Thus it seems that there are three distinct phases in this peculiar abnormal state of the generative functions.

"I have also noticed that, in most cases where the male plumage is in excess of the female, the tail-feathers are particularly long, some being as much as 19 inches in length.

"Although Mr. Yarrell states that this condition of the female generative organs is not confined to the *Phasianidæ*, and that it has occurred in the gold and silver pheasants, partridges, pea-fowls, common-fowl, common pigeon, kingfisher, the common duck, and that other classes of animals are liable to an influence similar in kind, particularly among insects and Crustacea, yet this disorganization is rarely observed except among the *Phasianidæ*, and particularly when these birds are produced in a domestic state, *i.e.*, on the present system of breeding pheasants in preserves. Very few *battues* take place in which some of these birds (generally designated males) are not killed and mixed indiscriminately with the heaps of the slain.

"As to the cause of this disorganization, if it occurred only in the old female, or if it were a common occurrence among birds either of different genera or of the same genus, it could be easily accounted for; but when it is generally found existing among a class of birds which are bred in vast numbers in a particularly artificial manner, it leads one to suppose that the cause must be connected with this condition."

In regard to all sorts of atrophy of the ovary it should be said that there is no known way to treat them. Such cases when they appear must be accepted by the poultryman as one of the vicissitudes of the business.

Gangrene of the Ovary

Salmon and other writers on poultry diseases following him have designated as gangrene a condition of the ovary relatively often found at post-mortem. Salmon's discussion of the matter is as follows: "This disease is quite common with all varieties of poultry. On examination of the ovary after death, the ova are found in different stages of development, but instead of being yellowish-pink in color, with the blood vessels well defined, they are brown or black, easily crushed and the contents broken down into a putrid liquid. Death is caused partly by peritonitis and partly by the absorption of the products of decomposition.

"The cause of this trouble is not well understood. It has been attributed to the birds being too fat, thus compressing the ovary and hindering the evolution of the ova. As it may occur in birds which are not fat and as it is evidently accompanied by the penetration and multiplication of bacteria, it is possibly an infectious disease."

We have not been able to find anywhere in the literature that there has been a thorough investigation of this disease.

Ovarian Tumors

Tumors and cancerous growths on the ovary are not uncommon. These include several sorts of interest to the pathologist, but not to the practical poultryman. From the literature it appears that at least the following (and probably other) kinds of new growths are found to occur on the ovary with greater or less frequency.

1. Benign tumors, of several types, including yolk tumors.
2. Carsinoma.
3. Dermoid cysts.

For a further discussion of the general question of malignant new growths see Chapter XX.

It is quite clear that none of these conditions can be successfully treated by the poultryman. In the first place any cancerous condition of the ovary or oviduct will practically never be diagnosed until after the bird's death. In the second place if it were diagnosed surgery would offer the only possible means of relief, and operations on the ovary are much too formidable for any one but the expert to undertake.

Abortion of Eggs

Regarding this matter Wright [1] has the following to say: "This is not to be confounded with the laying of soft eggs. These last are laid when mature, and usually by fat birds; but when violently driven or startled, or subject to violence of any kind, or even if suddenly and greatly terrified, immature yolks are sometimes detached from the ovary and expelled. This is most likely to happen with pullets not yet laying but about to lay, and being a real miscarriage or abortion, may wreck the constitution of a valuable bird

---

[1] Wright, L., "The New Book of Poultry." London, 1902, p. 574.

unless attended to. It is distinguished from the other by not occurring as a rule in fat birds; by the immature and small size of the yolk or yolks; generally also by hemorrhage; and always by signs of illness of chicks afterwards. Any such bird should be placed for a few days in a quiet and comfortable but rather dark pen, with a nest in case of need, and fed on a little bread and milk. Quiet rest is the main thing, but 20 grains bromide of potassium may be dissolved in half a pint of drinking water. With such care the event may be entirely recovered from."

Yolk Hypertrophy

There are a number of cases on record where the yolks formed by the ovary have been very much larger than normal. These "giant yolks" are due to a diseased condition of the organ, possibly contingent upon too much forcing for egg production. Such cases have been described by Gurlt,[1] and more recently by von Durski.[2]

When yolks become very large in this way they may break loose from the ovary without any rupture of the follicle wall along the stigma but a breaking or tearing loose of the stalk or pedicle of the follicle.

Failure of Follicle Wall to Rupture

Closely connected with the last diseased condition is one discussed by von Durski in which the follicle wall fails to rupture and release the yolk. In consequence of this, in the case described by von Durski, the follicle wall became stretched and pulled out into a long and very much twisted

---

[1] Gurlt, *Mag. f. d. ges. Tierheilk.* 1849.

[2] von Durski, "Die pathologische Veränderungen des Eies und Eileiters bei den Vogeln." Berlin, 1907.

stalk. This stalk held the hard and decayed yolk fast to the ovary. In cases of this kind the stalk sometimes breaks, and the yolk inclosed in the follicle and with the end of the stalk attached, passes down the oviduct acquiring albumen, membranes and shell. In still other instances the stalk breaks and the follicle and contained yolk drops into the abdominal cavity.

DISEASES OF THE OVIDUCT

Diseases of the oviduct are relatively common and cause a steady and probably in the aggregate rather large loss to the poultryman. Fortunately some of the diseases of the oviduct are more amenable to treatment than are those of the ovary. Further these diseases in many cases show plain external symptoms at a relatively early stage. Then they may be recognized and treated while it is still possible to effect a cure. This is usually not the case with ovarian diseases.

The general external symptoms of the commoner diseases of the oviduct are very much like those of constipation. The poultryman watching his birds is indeed rather likely to confuse the two. But if so no harm is done. The thorough cleaning out of the alimentary tract, and stimulation of the liver indicated in the treatment of constipation is the very best thing to be done in cases of inflammation and similar disorders of the oviduct.

Anatomy of the Oviduct

In order to understand more clearly the pathological conditions of the oviduct it is well to consider briefly at this point some facts regarding the normal anatomy and histology of this organ. Here we shall follow the accounts given by two of the authors in earlier papers.[1]

The oviduct of a laying hen is a large, much coiled tube

---

[1] Curtis, M. R., "The Ligaments of the Oviduct of the Domestic

filling a large part of the left half of the abdominal cavity. It is suspended from the dorsal body wall and lies dorsal to the abdominal air sac. Its anterior end is expanded into a large funnel which is spread out beneath the ovary in such a way that the mouth of the tube faces the ovary.

The oviduct is divided into five main parts, readily distinguishable by gross observation. Beginning at the anterior end, there are, in order: (a) the infundibulum or funnel (Fig. 57, A); (b) the albumen secreting portion (Fig. 57, B); (c) the isthmus (Fig. 57, C); (d) the uterus or "shell gland" so called (D); and (e) the vagina (E). The functions of these different parts have already been indicated (cf. pp. 248).

The chief features in the finer structure of the oviduct may be described as follows:

Two muscular layers, an outer longitudinal and an inner circular layer can be distinguished in all parts of the oviduct. The inner surface of the oviduct is thrown into a number of primary longitudinal ridges. The epithelium over these ridges forms secondary folds. In the uterus the ridges as such are lost, and instead there are a number of leaf-like folds of the inner surface.

Three types of glands are described: (1) Unicellular epithelial glands occurring between the ciliated cells in all parts of the oviduct except the anterior portion of the funnel. (2) Glandular grooves. These are accumulations of gland cells at the bottom of the grooves between the secondary folds of the epithelium. These are found only in the funnel region. But there they occur well towards the anterior end. (3) In all parts of the oviduct between the funnel and the vagina there is a thick layer of glands beneath the epithe-

Fowl." Me. Agr. Expt. Stat. Bul. 167, pp. 1–20, 4 plates, 1910.
Surface, F. M., "The Histology of the Oviduct of the Domestic Hen." Me. Agr. Expt. Stat. Ann. Rept., 1912, pp. 395–430, 5 plates.

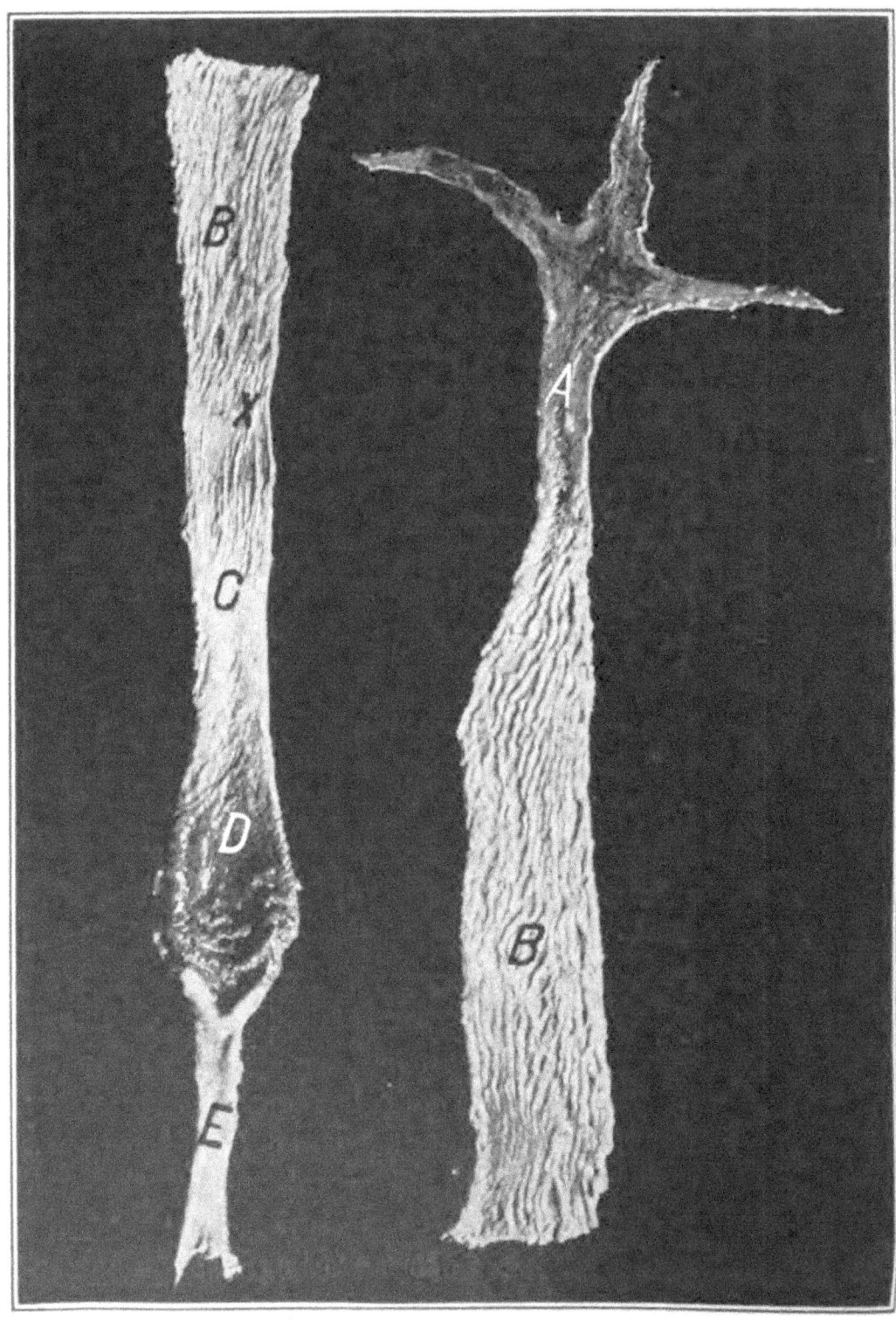

Fig. 57. — Oviduct removed from a laying bird and cut open along the point of attachment of the ventral ligament. It is opened back, showing the characteristic glandular regions. *A*, funnel; *B*, albumen-secreting region; *X*, isthmus ring; *C*, isthmus; *D*, shell gland; and *E*, vagina. (Original.)

lium. These are called *tubular glands*. They consist of long convoluted and branched tubules which open to the lumen of the oviduct by short epithelial ducts. These tubular glands are homologous, structurally at least, with the glandular grooves of the funnel. The tubular glands reach their greatest development in the albumen secreting region. Histologically the unicellular epithelial glands present a similar appearance in all parts of the oviduct except the vagina. In this latter region the cells are longer and much narrower and have a slightly different arrangement than in other parts of the oviduct.

The walls of the tubular glands consist of large gland cells which in the albumen portion and the isthmus of a laying hen have small, irregularly shaped, dark staining nuclei which lie well towards the basal ends of the cells. In these two regions the protoplasm of the cells contains rather coarse granules which vary greatly in size.

The line of demarcation between the albumen region and the isthmus is characterized by the absence of these tubular glands in that region. The cells of the tubular glands in the albumen region and in the isthmus present the same histological appearance.

In the uterus the cells which form the tubular glands have a somewhat different appearance. The nuclei of these cells are large with regular outlines and are situated near the center of the cells. The protoplasm is very finely granular and is quite different from the coarsely granular condition found in other parts of the oviduct.

The tubular glands or any homologous structures are entirely absent from the vagina. Only the unicellular epithelial glands occur there.

Inflammation of Oviduct

This is one of the most important and common diseases of the oviduct. It may occur alone or in association with other morbid conditions of this organ.

Diagnosis. — Combining the accounts of various observers, it may be said that a bird affected with inflammation of the oviduct at first shows indications of a desire to lay without being able to produce eggs, or it may lay eggs containing more or less blood or eggs without shells or small and misshaped eggs containing albumen but no yolk, or finally the yolk may be dropped without any covering of albumen or shell. There is a continual and violent straining (sometimes resulting in apoplexy). The wings are dropped and the feathers puffed out. As the inflammation increases there is high temperature, straining and an effort to rub the abdomen upon the ground. In later stages the bird becomes dull, indisposed to move and the comb is pale.

Etiology. — There are probably to be distinguished three classes of causes which lead to inflammation of the oviduct. These are:

1. Physiological; from irritation due to too frequent laying or from too stimulating foods or condiments.

2. Traumatic; from irritation due to too large eggs, or to the breaking of eggs within the oviduct, or to similar causes.

3. Specific infection; it is probably that alone or in combination with the causes classed under 1 and 2 a specific infection of the lining membranes of the oviduct may occur.

In an inflamed oviduct there very often is a copious serofibrinous exudate. This hardens about any foreign body (egg, broken egg, etc.) which may be in the oviduct, and by accretion causes this foreign body to increase in size. This, of course, makes it still more irritating, which in turn provokes further inflammation of the walls of the duct. One sometimes finds relatively enormous masses of material in a diseased oviduct, which have been built up in this way. There is an extensive literature on these "egg concrements" or "yolk tumors" built up either in the oviduct or in the abdominal cavity by hardened fibrous exudate, about an original basis of a broken, or miscarried, or aborted yolk or yolks. It is not necessary to review this literature here as it is only of interest to the specialist.

Treatment. — If this disease is to be dealt with at all the treatment must be individual, since it is something which will never affect considerable numbers of the flock at the same time. If individual treatment is to be successful it must be begun at a relatively early stage of the disease. Therefore, it is important that a bird showing the symptoms which have been described above should be isolated at once and as a first step in the treatment given a purgative dose of Epsom salts (see p. 53). All stimulating foods such as meat, green cut bone, linseed meal and similar substances, as well as condiments like condition powders, pepper, etc., should be immediately taken away from the bird. A light ration and plenty of green food should be given. Salmon recommends following the purgative with $\frac{1}{2}$ drop of tincture of aconite root 3 times a day. Equally effective, and much easier to administer, will be found 1–10 gr. aconite root tablets (see p. 55).

Prolapse of the Oviduct (Eversion)

It not infrequently happens, from one cause or another, that the lower portion of the oviduct becomes everted and projects from the vent as a mass of red or purplish tissue. This condition is known as *prolapsus* of the oviduct.

Diagnosis. — The diagnosis of this diseased condition is simple and consists merely in the observation of the prolapsed oviduct. If there is a mass of red or bloody tissue projecting from the vent, one is safe in diagnosing prolapsus. The only point which needs particular attention in the diagnosis is as to the degree to which prolapsus has occurred when the bird is discovered. The importance of this lies in the fact that on it depends the treatment which it is advisable to give. Where the prolapse is only partial and is discovered early it is advisable to treat it by the methods outlined below. If, on the other hand, the prolapse is extensive and has existed for some time before the bird is seen so that the mass of tissue has turned a blue or purplish color or has been pretty extensively picked and torn by the other birds in the pen, then it is useless to carry on any treatment and the proper thing to do is to kill the bird at once.

Etiology. — Prolapse of the oviduct may be caused by a number of different things. It is observed not only in old hens, but, in our experience, quite as frequently in pullets. The fundamental cause of the condition is, of course, a weakness of the oviduct walls and ligaments, chiefly in respect to their muscular portions, which makes the oviduct unable to stand the strains put upon it in egg production. The immediate cause may be either:

1. Straining to lay a very large (double yolked) egg. This is perhaps the most common cause.

2. Straining to lay when there is an obstruction in the oviduct (egg bound).

3. Constipation. The rectum full of hardened feces stimulates all organs in that region of the body to expulsive reflexes.

4. Zürn says that oftentimes feces may become lodged in the cloaca in a sort of blind pocket, and then set up the same expulsive reflexes as an egg in the cloacal or vaginal regions normally does. In the effort to expel this foreign body the oviduct may become everted.

Diseases of the Reproductive Organs

The most serious thing about prolapsus is that if not discovered very shortly after it occurs it is almost sure to result fatally, because the everted portion will become so badly infected as to cause blood poisoning, or the protruding mass of tissue will be picked and torn by the other birds in the pen until there is no hope of repair, whatever the treatment.

Treatment. — As stated above, the advisability of treating prolapsus depends upon its degree and duration before discovery.

In treating this condition the first thing to endeavor to do is to remove the cause. That is, if the bird is constipated give it a rectal enema of warm soapy water, followed by $\frac{1}{2}$ teaspoon of Epsom salts by the mouth. If there is a lump of feces lodged in the cloaca this should be carefully removed. The protruding mass of tissue should be washed with warm 1 to 1000 bichloride of mercury solution, or a warm $\frac{1}{2}$ per cent cresol solution. After the protruding parts are thoroughly cleansed they should be well greased with vaseline, or with the ointment already recommended (p. 55). Then with the fingers well greased an effort should be made to replace the protruding mass in the body. In doing this one should proceed with the greatest gentleness. In most cases with care and patience it is possible to reduce the prolapsus, that is, to get the extruded tissue back into the body in approximately its normal position.

After the parts have been carefully replaced in normal position the next point to be considered in the treatment is to insure that they shall stay there. That is to say, it is necessary some way to bring about a healthy degree of contraction of the muscular walls of the oviduct so as to hold the parts in place permanently. In order to do this Salmon recommends the use of ergot. Robinson follows Salmon in this recommendation. It should be said, however, that it is doubtful whether this treatment is advisable.

Ergot is a rather violent poison for poultry. It seems likely that the treatment recommended by Salmon and Robinson is based on a theory that the action which ergot has on the mammalian uterus will be duplicated on the fowl's oviduct rather than upon actual experience in administering the drug to poultry. The measure recommended by Zürn to bring about a healthy contraction of the replaced oviduct in cases of prolapsus would seem to be simpler and on the whole more likely to yield desirable results than the ergot treatment. Zürn recommends that a lump of ice be placed in the cloaca after the prolapsed oviduct is returned to its place and that this treatment be followed up for some hours.

The bird should be kept in a small coop, partly darkened, where there will be every inducement for it to remain perfectly quiet. The success of the treatment depends very much on keeping the bird quiet for a few days. It should be fed only a light and unstimulating ration with plenty of green food.

Prognosis. — If discovered early enough prolapsus is curable.

Obstruction of the Oviduct ("Egg Bound")

Perhaps the commonest of all diseased conditions of the oviduct is that which leads the poultryman to say that a bird is "egg bound." By this is meant that there is something in the oviduct which the bird is not able to pass to the outside and which in turn prevents the normal passage of eggs. In many cases this is not properly speaking a disease at all but rather an accident. Other cases, however, depend upon a true diseased condition of the oviduct.

Diagnosis. — The symptoms of this trouble, as they are usually described, consist chiefly in the obvious fact that the hen is trying to lay but cannot extrude the egg. If this struggle is kept up long enough the bird will become exhausted, and show it by keeping quiet, with roughened plumage and the general aspect of being ill. Sometimes the egg can be felt from the vent.

All these general symptoms of egg bound condition may be observed in mild form in a great many cases with birds which subsequently lay the egg without trouble. In many instances the extrusion of an egg which is finally successfully laid is attended with a good deal of difficulty. There are all degrees of gradation between this somewhat difficult but still normal laying and the condition of complete obstruction of the oviduct where the egg cannot be passed at all. The practical consideration to which this leads is that one should not be too hasty in applying treatment for the egg-bound condition. A diagnosis of the trouble, in other words, should not be finally settled upon until there remains no doubt that the hen is not going to pass the egg without help from the outside.

It must also be remembered that in many cases of obstruction of the oviduct, the obstruction is so far up that it cannot be felt from the outside. In such cases the diagnosis must be made upon the general behavior of the hen, and in particular in regard to going frequently on the nest without laying.

Etiology. — In considering the causes of obstruction of the oviduct it is necessary to distinguish between several different sorts or categories. This may be done as follows:

1. Simple "egg bound" condition, in which a normal egg is lodged in the uterus or vagina and cannot be expelled. This inability to expel the egg may be due to any one or a combination of the following causes acting together:

a. Egg of too large size, so that it is mechanically difficult or impossible to force it through the natural passage. Robinson regards this as the most common cause.

b. Exhaustion (true physiological fatigue) of the muscular walls of the oviduct. This condition results after long continued and unsuccessful attempts to expel the egg. It leads to

c. Atony and paralysis of the duct, in which the muscular walls are incapable of making any effective contraction at all.

2. Complicated "egg bound" conditions in which the fundamental source of the trouble is not simply mechanical, and in which usually the portions of the oviduct anterior to the uterus are involved. In this general category the following sorts of cases are to be included.

a. Atony and paralysis of the upper portions of the oviduct. This condition may exist for a long time without being recognized.

b. Inflammation of the oviduct leading to the formation of fibrous exudate which accumulates in the duct, until it may form a mass of relatively enormous size (usually with one or more yolks as a nucleus) completely obstructing the duct, and eventually leading either to gangrene or rupture of the walls, or both.

c. Volvolus, or twisting of the oviduct about its own long axis, completely obliterating the cavity.

d. Stenosis or stricture of the oviduct. This may result from several causes. One frequent one is that in laying a very large egg the oviduct wall becomes torn to greater or less degree, and subsequently heals. The scar tissue contracts the cavity and a stricture is thus caused.

Treatment. — Whether treatment is or is not likely to be effective depends upon which of the two main categories above defined any given case belongs to. Simple obstruction of the oviduct may be successfully treated. In cases of complicated obstruction treatment is not indicated, for a variety of reasons. These conditions are in the first place

Diseases of the Reproductive Organs

difficult to diagnose, and offer little prospect of successful cure even after a diagnosis has been made.

The best advice which has come to our attention for the treatment of the simple egg bound condition was published some years ago in the English journal *Poultry* and is here quoted verbatim:

"It is a good plan to watch those birds that are about to lay. Should they visit the nest frequently during the course of the day and leave without depositing an egg, it is almost certain that something is wrong and when a pullet is in such a state there are three good remedies that may be tried. The first is: Take the bird up gently, and hold her so that her stern is over the mouth of a jug of boiling water, that the steam arising therefrom may get to the parts and help to relax and procure delivery of the egg. If this has not the desired effect after an hour's rest in a quiet coop, the vent should be oiled gently with a feather, and the hen given a powder composed of 1 grain of calomel and 1-12 grain of tartar emetic. The powder may be mixed in a bolus of food, and put into the bird's crop. If it be acting properly, a marked improvement should be noticeable in the bird a few hours afterwards, while a second powder given two days subsequently will probably complete the cure. It is advisable for a while to feed the fowl sparingly on a somewhat low diet, withholding any fat forming food, and giving lime-water to drink, after the system is rid of the powder. The second remedy was advocated by Dr. H. B. Greene, . . . and is best applied when the egg can be felt. It is: Let an assistant, seated on a chair, hold the bird firmly on his knees on its back, with the vent directed away from him. Seating yourself opposite, with the finger and thumb of the left hand outside the bird's body, push the egg firmly but carefully towards the vent, until it is plainly visible, and, keeping it in that position, with a bradawl in the right hand puncture the egg shell, evacuate the contents of the egg with an egg-spoon, and afterwards with a pair of tweezers break down and take out the shell piece by piece until assured by passing the finger into the vent, that the cloaca is empty. Special care must be taken to avoid injuring the bird with the point of the awl; and one's assistant must maintain a steady and firm hold on the fowl. A third method of relieving an egg bound hen was recommended by a correspondent in our issue of June 10, 1898, and has since been frequently tried by several poultry keepers, and found very efficacious. 'When a hen is in that state I hold her over some hot water, bathing the vent at the same time. After this I use a small penknife (blunt) in the following manner: Placing the edge of the blade along the first finger so that the end is level with the finger end, I push the finger with the knife into the vent until they touch the egg; then I begin to scrape until I hear that I have scraped the rind or skin away from the egg (I mean outside the egg). The hen is then placed on the nest, and I will guarantee she will lay in 20 minutes, or in most cases even less than that. I got this advice from a man who has kept poultry on a small scale for 50 years. I have tried it several times, and have never known a hen to be egg bound a second time. This method, it would appear, saves the egg. The great thing throughout is to keep the bird quiet, and in future to avoid extra fat forming food.'"

Prognosis. — Good in cases of simple obstruction if taken in hand early; bad in all cases of complicated obstruction.

Rupture of the Oviduct

In some cases of complicated obstruction, and in cases of severe inflammation the walls of the oviduct may break and allow the contents to escape into the abdominal cavity.

In such cases death usually ensues in a relatively short time as a result of peritonitis. These cases are incurable; indeed the trouble is usually not known till after the bird dies. The lower portion of the oviduct (vagina) or the cloaca may be ruptured in passing a very large egg. If the wounds made in this way are relatively small they will usually heal without any trouble. If, on the other hand, such tears are extensive they may very easily become infected, and unless treated properly in accordance with the general directions given in Chapter XXI for the treatment of wounds, the bird will die of blood poisoning.

Lately in some experimental work we have shown that extensive ruptures of the upper part of the oviduct (albumen portion) will heal spontaneously and leave no trace of the injury. We have removed large pieces from the oviduct wall in this way, only to find the bird laying in a perfectly normal way in a month or so after the operation. At autopsy only a small scar or else no trace whatever of the wound could be found.

Gangrene of Oviduct

This may result from severe and complicated obstruction. What is meant by "gangrene" is that walls of the oviduct die, and putrefy. This causes general blood poisoning from which the bird dies. Gangrene of the oviduct most frequently follows severe cases of complicated obstruction where there is a mass of fibrous exudate deposited in the oviduct. There is not the slightest hope of successfully treating such cases.

Breaking of Egg in Oviduct

It sometimes happens that an egg in the upper portion of the oviduct, before it has acquired any shell, is by accident broken. There is a belief common amongst poultrymen that this is always immediately fatal. There is but little discussion of the subject in the literature, but our experience here indicates that two sorts of results may follow the breaking of an egg in the oviduct. These are:

1. An inflammatory condition of the oviduct is induced leading to copious secretion from the glands of the albumen portion of the duct and the isthmus. There is also a copious fibrous exudate, and the final outcome is a severe case of complicated obstruction of the oviduct. Death in these cases may be delayed for a long time after the original accident. In the absence of inflammation recovery may possibly occur.

2. Death within a short time (2 to 3 hours) after the breaking of the egg, without visible lesion of any organ of the body. The oviduct is not even inflamed. Absolutely the only things which are not normal in such cases are (a) the broken egg in the oviduct, and (b) the fact that the bird is dead. We have had several such cases come to autopsy. They are very puzzling. In them is to be found the basis for the poultryman's belief as to the fatal character of this accident. In reality it seems probable that in these cases the thing which caused the egg to be broken was also the cause of the death of the bird. That is, a blow, or any sort of sudden shock violent enough to break an egg in the oviduct might also very well be the cause of death. Such cases need further study.

Abnormal Eggs

Owing to various diseased conditions of the oviduct many different kinds of abnormal eggs are produced by fowls. The explanation of the different types of such eggs is usually tolerably clear if one gets definitely in his mind the normal physiology of egg production as outlined above. We shall consider here only some of the more important general classes of such abnormal eggs. Such eggs are very interesting from the scientific standpoint but are of relatively little practical significance to the poultry keeper because of the rarity of their occurrence.

Soft-shelled Eggs. — These are eggs laid without a sufficient amount of shell substance covering the shell membrane. The immediate cause lies in a failure of the uterus to function properly. Regarding this class of abnormal eggs Wright has the following to say: "Soft eggs may be caused by lack of shell-material, which, if discovered, points to the remedy, the most rapid being pounded raw oyster-shell. Or they may be caused by the fowls being driven or frightened, in which case they soon cease, and nothing need be done unless the injury has been so severe as to prematurely detach small and unripe yolks, when the case becomes a real *abortion*, or they may be caused by condiments and too much animal food, spices in particular leading frequently to all sorts of trouble with the egg-organs, particularly in the Mediterranean races of poultry. A few small doses of Epsom salts or jalap, and cessation of the extra stimulus, will remedy this. But far the most usual cause is simple over-feeding. A little careful investigation will find which is in fault, and that will indicate the appropriate remedy. Want of shell material is far less common than it used to be; over-feeding or over-stimulation probably more so."

Small, Yolkless Eggs. — These little eggs, variously called "wind-eggs," "cock eggs," "witch eggs," "luck eggs," etc., are familiar to every poultry keeper. They contain no definitely formed yolk, and to the casual observer seem to consist of nothing but a small shell filled with white. The laying of one of these eggs is popularly supposed to mark the end of a laying period. This belief is without foundation in fact. They may be produced at any time. Un-

T

published data collected over a period of years at this Station in regard to such eggs indicate that three factors are fundamentally concerned in their production. These are:

1. The bird must be in an active laying condition; the more pronounced the degree of physiological activity of the oviduct, the more likely are these eggs to be produced.

2. There must be some foreign body, however minute, to serve as the stimulus which shall start the albumen glands secreting. This foreign body may be either a minute piece of hardened albumen, a bit of coagulated blood, a small piece of yolk which has escaped from a ruptured yolk, etc.

3. It seems likely, though this is a point not yet definitely settled, that ovulation (*i.e.*, the separation of a yolk from the ovary) must precede the secretion of albumen around the foreign body to form one of these eggs.

Double and Triple Yolked Eggs. — Eggs with two yolks are, of course, quite common. They result from a disturbance of the time relations of ovulation, of such nature that two yolks get into the oviduct at nearly the same time and become surrounded by common layers of albumen.

Eggs with three yolks are very rare. An egg of this kind laid by a pullet at this Station is shown in Fig. 58.

Studies made by one of the authors [1] have thrown considerable light on the general problem of the cause and nature of multiple-yolked eggs. Summarized some of the chief results were as follows: As to frequency it appears that the Maine Station flock, over a long period of time, produces 531 single-yolked eggs to every double-yolked egg. That is,

---

[1] Curtis, M. R., "Studies on the Physiology of Reproduction in the Domestic Fowl." VI. Double- and Triple-Yolked Eggs. *Biol. Bul.*, Vol. XXVI, pp. 55–83, 1914.

——— "Studies on the Physiology of Reproduction in the Domestic Fowl." XI. Relation of Simultaneous Ovulation to the Production of Double-Yolked Eggs. *Jour. Agr. Research*, Vol. III, pp. 375–385, 1915.

Diseases of the Reproductive Organs

only two tenths of 1 per cent of the eggs are double-yolked. The ratio of double to single yolked eggs is less than twice as high as the ratio of twin to single births in the human family.

All birds are not equally likely to lay double-yolked eggs. In fact the great majority of birds never lay anything but single-yolked eggs. There are, however, birds which possess a tendency to lay double-yolked eggs. Such an individual may produce several such eggs. It has been further found

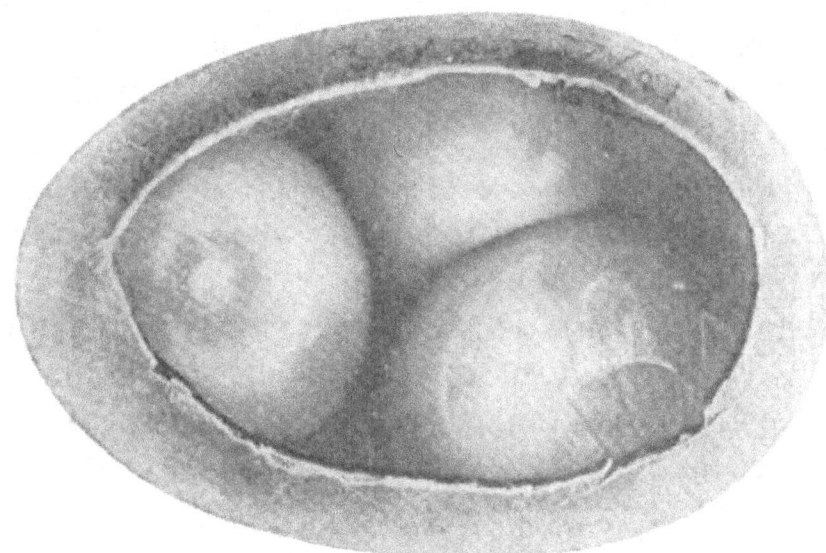

Fig. 58. — Triple-yolked egg. (Original.)

that a bird which possesses the tendency to lay double-yolked eggs is not equally likely to produce them at any age. She is most likely to produce them when she is young. Eighty per cent of all the double-yolked eggs produced by the Station flock are produced by birds less than eight months old. We have only a very few records of birds which have laid double-yolked eggs after their first adult molt.

It has been usually supposed that double-yolked eggs are caused by the simultaneous entrance of two yolks into the egg tube and the consequent common passage of the two yolks through the duct. A careful study of the structure of all the double-yolked eggs produced by the Station flock shows that in only a very small per cent of the cases (about 16 per cent) have the two yolks passed the entire length of the duct together. In such cases the two yolks are inclosed in a common thin layer of white membrane, the chalazal membrane, and have only one pair of chalazæ. They also have common albumen envelopes as well as common egg membrane and shell.

Since the formation of each egg part (chalazal membrane and chalazæ, thick albumen egg membrane, and shell) is confined to a particular part of the oviduct, a study of the number of secondary parts which are common to the two yolks of a double-yolked egg shows the level of the duct where the two yolks came together. Such a study carried out on all double-yolked eggs produced by the large flock of birds owned by this Station shows that the two yolks unite at every level of the duct from the mouth of the funnel to the very end of the albumen secreting portion. It shows further that the number of eggs of any given structure observed is exactly equal to the number expected on the assumption that the union of the two yolks occurs indiscriminately at every level of the duct from the mouth of the funnel to the beginning of the isthmus or egg membrane secreting portion. When two eggs unite after the first egg has received its membrane the result is two eggs at the same time.

The structure of the egg has shown us that in a majority of cases the two yolks of a double-yolked egg have not passed the entire length of the duct together. On a moment's reflection we see that there was never any *a priori* reason for the assumption that the cause for the production of a double-yolked egg was necessarily the simultaneous discharge of two yolks from the ovary into the oviduct or egg tube. The only condition necessary for two yolks to be inclosed in the same egg membrane is that they enter the membrane secreting portion of the oviduct together. There are at least three possibilities besides simultaneous ovulation which may bring two yolks together before they reach this portion of the oviduct. First, the first yolk may be delayed at any level of the duct forward to the point where the egg membrane begins to be secreted; second, the first yolk may be returned up the oviduct and then come back in company with the second yolk; and third, a yolk may be ovulated into the body cavity and picked up by the oviduct shortly before or after the ovulation of another yolk. It is, therefore, unnecessary to assume that the production of a double-yolked egg represents simultaneous or even an abnormally rapid succession of ovulations, since any of these delays may have been as long as the normal period between ovulations.

A study of the structure of the eggs and the egg records of the birds leads to the conclusion that double-yolked eggs do not necessarily represent two simultaneous or even nearly simultaneous ovulations; but in about a third of the cases of double-yolked eggs the time between the two ovulations must have been unusually short since the birds which laid these double-yolked eggs each laid a normal egg on the preceding day. A study of the egg structure of these double-yolked eggs, where the time between the ovulations is known to have been abnormally short, shows that they have been simultaneous in only a small per cent of the cases. In fact the two yolks have come together at every level of the duct in front of the beginning of the isthmus.

A study of the ovaries of birds which had recently produced double-yolked eggs showed that each of the two yolks was discharged from a normal separate follicle exactly as are the yolks of successive single-yolked eggs.

From these recent studies of double-yolked egg production it is certain that some individual hens have an inherent tendency to lay double-yolked eggs, while a great majority of hens never lay anything but normal single-yolked eggs. A bird with the tendency to double-yolked egg production is more likely to produce double-yolked eggs when she is quite young than later in her life.

The two yolks of a double-yolked egg may enter the oviduct simultaneously and pass the entire length of the duct together receiving an entire common set of egg envelopes, or they may come together at any level of the oviduct from the funnel mouth to the beginning of the isthmus. 'It is highly probable that the two ovulations may be either simultaneous or that they may be separated by any period up to the normal time which elapses between ovulations.

The production of a double-yolked egg is evidently seldom caused by the simultaneous discharge of two normal separate follicles into the oviduct. More often it is caused by the successive discharge of separate follicles at times varying from simultaneity to the normal period and by the subsequent union of the eggs in the duct due to a difference in the rate of passage of the successive eggs.

Inclusion in Eggs. — The number of different foreign substances which at one time or another have been found inclosed in eggs is great. The list includes blood streaks or spots, blood clots of firm consistency and often considerable size, lumps of bacteria, worms, fecal matter, etc., etc.

From the practical standpoint the only inclusions which need consideration are blood spots. Many inquiries are annually received at this Station as to what causes these spots and what to do to get eggs which will be free from them. These inquiries are most frequent in the spring months. *The only thing which can be done in such cases is to candle the eggs and sell only those which show no spots.* Hens which are

Diseases of the Reproductive Organs 279

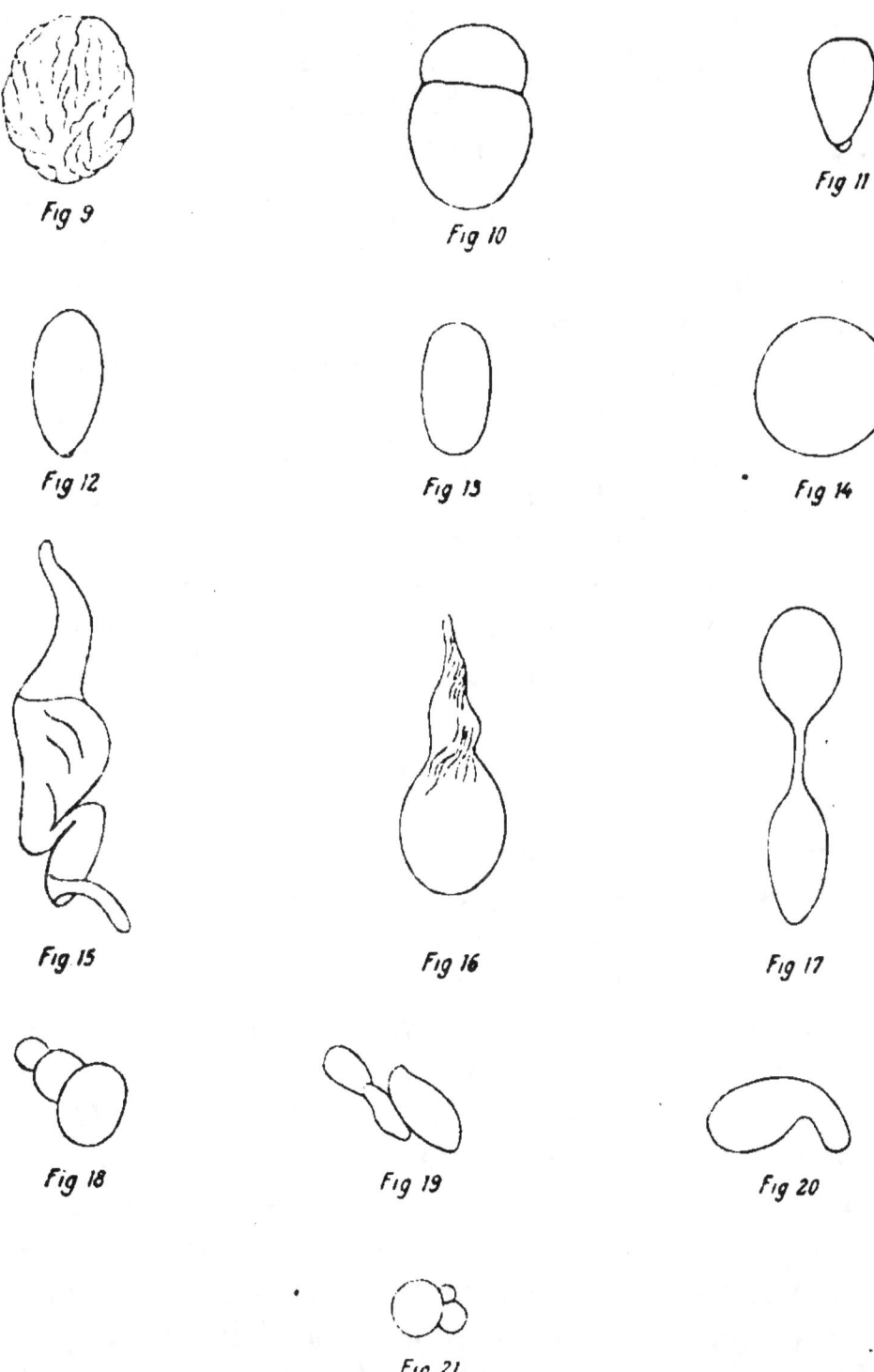

Fig. 59. — Showing shapes of abnormal eggs sometimes found. (From von Durski after Landois.)

perfectly normal often lay eggs with blood spots, especially in the spring of the year when laying is heavy. The blood which makes the spot probably comes in most cases from the ovarian follicle. When this ruptures a little blood escapes into the oviduct and is caught up in the albumen. The so-called "liver" or "meat" spots in eggs are in nearly every case thoroughly hardened, well packed together, blood clots. They may be of large size. These inclusions *do not* represent, as they are sometimes said to, portions of the oviduct wall which have been torn off and inclosed in the egg.

Eggs of Abnormal Shape. — There are many other kinds of abnormal eggs besides those here discussed, but as they have no practical significance it is not desirable to devote further space to them. In closing this section we append some figures showing in outline some of the curiously shaped eggs which have been found.

Vent Gleet (Cloacitis)

This is a true venereal disease of poultry. It usually begins with a hen, but is transmitted in copulation to the male, and by him to other birds in the flock.

Diagnosis. — Salmon gives the following clear account of the symptoms: "The first symptom observed is the frequent passage of excrement which is voided in small quantities almost as rapidly as it reaches the cloaca. Often the bird endeavors to drop excrement when cloaca is entirely empty. This action is due to the tenderness and irritability of the cloaca which gives to the bird the sensation of fullness, and produces spasmodic contractions. If an examination is made the mucous membrane is found in the early stages to be red, dry, swollen and hot. In a day or two a discharge makes its appearance. It is, at first, thin and watery, but soon becomes white, purulent, and offensive. This discharge collects upon the skin and feathers about the vent, obstructs the passage and irritates the parts with which it comes in contact. The soiled skin becomes red and inflamed, it may be abraded by friction or by the bird picking at it, and thus sores or ulcers are started which may become quite troublesome."

Etiology. — The cause of the disease has not yet been thoroughly worked out. Wright suspected it to be identical with human gonorrhea because of the similarity of symptoms, infectiousness, etc. However, he has not been able to isolate the *Gonococcus*, or specific germ of gonorrhea from affected birds.

Lewis and Clark [1] report an outbreak of vent gleet among recently purchased Orpingtons at the poultry plant on the New Jersey college farm, as a result of which the average egg production for the year was but 25 per cent of the theoretical. The percentage of fertility of the eggs was low, about 60, although apparently good vigorous male birds were used. The disease was found to be very hard to overcome, and it is concluded that in dealing with it the best method lies in the destruction of the affected fowls.

Treatment. — The following is the treatment outlined by Wright: "Any hen found with it should at once be isolated, and the male bird carefully examined, and if necessary also isolated. Give 30 grains Epsom salts, and twice a day inject first a 4 per cent solution of cocaine, and immediately afterwards a solution of nitrate of silver 4 grains to the ounce. The fifth day commence a small copaiba capsule daily, and inject acetate of lead, 1 dram to the pint. Feed rather low meanwhile, and dust any sore places outside with iodoform or aristol. If not well after 2 or 3 weeks, we would kill the bird, as the disease is not quite free from danger;

---

[1] Lewis, H. R., and Clark, A. L., "Poultry Diseases." N. J. Agr. Expt. Stat. Rept., 1913, pp. 276–279.

for if the operator should touch his eyes accidentally before he has cleansed his hands, the result might be a most violent inflammation."

Diseases of the Male Reproductive Organs

A number of diseases of the male reproductive organs have been described, but they are all of no practical significance, for the reason that no poultryman ought ever to use as a breeder a male bird that ever had any disease of these organs, whether it had been "cured" or not.

CHAPTER XVIII

White Diarrhea

Of all the diseases which the poultryman is called upon to fight, there is probably none so destructive, year after year, as the disease (or diseases) known as "white diarrhea." The loss of chicks ascribed to this cause varies in different years and in different places from 10 to 90 per cent. It is perhaps not too much to say that more than 50 per cent of the chicks hatched throughout the country are lost from white diarrhea in its various forms. The number of inquiries concerning this disease which are annually received, and the amount of space devoted to it by the poultry press, lead one to believe that "white diarrhea" is perhaps the worst enemy with which the poultryman must contend.

White diarrhea is more common among artificially hatched and brooded chicks than among those which have been hatched and cared for by hens. However, it is by no means unknown among the latter. Many poultrymen report as heavy mortality from this disease among hen hatched and reared chicks as from those which were incubated and brooded by artificial methods.

Almost any chick that comes out of the shell apparently healthy on the 21st day will live for the first week. If white diarrhea is going to strike the brood they usually begin to show symptoms about the end of the first week. The heavy loss of chicks from this disease occurs between the ages of one and three weeks. Where the brood is badly affected chicks may continue to die until the fourth or fifth week. On the other hand if a brood goes through its first three weeks of life without being attacked by this disease it is practically safe from its ravages. White diarrhea then may be said to be limited to the first three weeks of the chick's life so far as serious mortality from it is concerned. The reason for this no doubt is that the digestive system of chicks under three weeks old is so delicate that even a slight disturbance makes a very serious handicap for the chick.

Etiology. — Within recent years a large number of studies concerning the cause, prevention and cure of white diarrhea have been conducted. Investigations have been carried on by state and national institutions as well as by many private individuals. Consequently a large number of alleged causes of the disease are given by different writers. Among these may be mentioned: Debilitated breeding stock, improper incubation, improper brooding, overheating, chilling, poor ventilation, over-crowding, poor or improper food and filth as well as specific bacteria, fungi or other parasitic organisms.

It is doubtful if many of the cases of true *white diarrhea* are caused by physical or mechanical agents. In most cases true white diarrhea appears to be an infectious disease. Such disease we know is caused by some form of parasitic organism. Without doubt improper incubation, brooding and feeding, resulting in weakened chicks, very often lay the foundation for the attacks of parasitic organisms. In many cases these faulty methods of handling the eggs and chicks appear to be the real cause of the disease while they are really only indirect causes.

From this it should not be understood that such things as poor food, poor brooding and weakened breeding stock are of no importance in the study of white diarrhea. It is just exactly these predisposing factors which result in chicks with weak constitutions, easily overcome by disease germs. Without doubt the points at which most progress can be made in combating such diseases are in the methods of incubation and in the care of the chicks for the first three weeks of their lives. Nevertheless it should not be forgotten that the death of the chick is caused by the ravages of some minute parasitic organism.

Within recent years several investigators have discovered organisms which they believe to be the specific cause of white diarrhea. Three of these may be mentioned at this place: (1) *Coccidium tenellum* or *cuniculi* producing the disease called "coccidiosis." (2) *Bacterium pullorum* producing "bacillary white diarrhea" and (3) *Aspergillus fumigatus* and allied species, producing aspergillosis or brooder pneumonia of chicks. Of these the first two diseases will be considered in some detail in the following paragraphs. Aspergillosis is treated in a separate section of this chapter (cf. p. 173).

Intestinal Coccidiosis

In 1908 Morse [1] published a preliminary account of some investigations on the cause of white diarrhea. He claimed that microscopic examination of the intestines of chicks, dying with this disease revealed the presence of large numbers of protozoan organisms which he identified as *Coccidium tenellum*. Cole and Hadley [2] of the Rhode Island Experiment Station reported finding a similar organism in white diarrhea chicks. They identified it as *Coccidium cuniculi*. These two species of coccidium are so nearly alike that it is

[1] Morse, G. B., "White Diarrhea of Chicks." U. S. Dept. of Agr. Bur. Anim. Indus. Circ. 128, pp. 1–8, 1908.
[2] Cole, L. J., and Hadley, P. B., "Blackhead in Turkeys." Rhode Island Agr. Expt. Stat. Bul. No. 141, pp. 138–272, 1910.

very difficult to distinguish them except by prolonged study of their life cycles.

Various species of coccidia have long been known to infest many domestic animals. A number of these have been described in fowls and other birds. In many cases they produce very serious lesions. But the contention of Smith [1] that in many cases these parasites are more or less normal inhabitants of the digestive tract seems fairly well founded.

Coccidiosis in birds and the relation of coccidia to the disease known as entero-hepatitis have been discussed on pages 71 and 94. On page 73 a detailed description of the life history of a typical coccidium is given. It is supposed that it is the same coccidium causing entero-hepatitis in turkeys which is related to white diarrhea. For further description of this organism the reader is referred to the preceding chapter.

Hadley and Kirkpatrick [1] have reported some feeding experiments with these coccidia in which they have been able to produce the disease in chicks several days or weeks old. As will be noted later, it is claimed that infection with the bacterium of bacillary white diarrhea must take place during the first two or three days of the chick's life if it is to produce the disease. It is possible that these facts may be of some use in distinguishing the two forms of this disease.

Diagnosis. — The symptoms of coccidiosis are similar to those of other forms of white diarrhea (cf. p. 292). The only exception is that according to Morse the ceca are always distended with yellowish-white cheesy matter. In other

---

[1] Smith, Theobald, "*Amœba meleagris.*" *Science*, N. S., Vol. 32, pp. 509–512, 1910.

[2] Hadley, P. B., and Kirkpatrick, W. F., "Further Investigations upon White Diarrhea of Chicks." *Successful Poultry Jour.*, Vol. 14, pp. 18–19, 1909.

forms of white diarrhea this may or may not be the case. These different forms of white diarrhea have been too little studied as yet to permit of an exact differential diagnosis on external symptoms even supposing that ever to be possible.

Up to the time of writing no further work has appeared to substantiate the claims that this coccidium is an important cause of white diarrhea. On the other hand, work with the bacillary form of this disease has been carried on by a number of investigators. At the present time it appears that by far the greater number of epidemics of white diarrhea are caused by the bacterium described below.

Bacillary White Diarrhea

In May, 1908, Rettger and Harvey[1] published a paper on "Fatal Septicemia in Young Chickens or White Diarrhea." From a large number of observations and experiments they came to the conclusion that white diarrhea was caused by a bacterium. A number of later papers by Rettger and his associates have appeared since then. In these it has been clearly proven that at least one form of white diarrhea is caused by a bacterium.

Rettger took chicks which had died with all the symptoms of white diarrhea and by the ordinary bacteriological methods obtained pure cultures of a bacterium which had certain definite reactions and habits of growth. By these methods this bacterium can be distinguished from other kinds. To this species of bacteria he gives the name *Bacterium pullorum*. If entirely healthy chicks are inoculated with the pure culture of this bacterium they almost invariably show

---

[1] Rettger, L. F., and Harvey, S. C., "Fatal Septicemia in Young Chickens or White Diarrhea." *Jour. Med. Research.*, Vol. 18, pp. 277–290, 1908.

symptoms of white diarrhea and in many cases die. To cite only one case; at the Storrs Experiment Station (Bul. 68) chicks 12 hours old were fed cultures of this organism. During twenty-five days allotted to the experiment the mortality of the infected chicks was 76 per cent, while that of the controls was only 4 per cent.

In many cases Rettger was able to find *Bacterium pullorum* pure in the artificially infected birds. Further he was able to obtain the same bacterium from a large number of different chicks gathered from widely different localities.

These results of Rettger and his associates have been confirmed not only by their own later work but also by others. Thus Jones [1] was able to produce the disease in healthy chicks by inoculation with this organism. Inoculation of chicks 24 hours old gave a mortality of 82.5 per cent, while in the controls it was only 2 per cent. Gage [2] isolated this organism from the ovaries of adult fowls and proved by inoculation that it would produce the disease in young chicks.

Conclusive evidence has been produced that it is only during the first few days of its life that a chick is liable to infection with this disease. It has been shown at the Storrs Station [3] that the greatest danger of infection lies within the first 48 hours, but that infection may take place up to four days or occasionally later in the case of weak chickens. The same writers have shown that in the majority

---

[1] Jones, F. S., "Fatal Septicemia or Bacillary White Diarrhea of Young Chickens." Report of N. Y. State Vet. College for 1910, pp. 111–129.

[2] Gage, G. E., "Notes on Ovarian Infection with *Bacterium pullorum* (Rettger) in the Domestic Fowl." *Jour. Med. Research*, Vol. 24, pp. 491–496, 1911.

[3] Rettger, L. F., Kirkpatrick, W. L., and Stoneburn, F. H., "Bacillary White Diarrhea in Young Chicks." Conn. (Storrs) Agr. Expt. Stat. Bul. 74, pp. 155–185, 1912.

of cases the hen is the source of the infection. It has been fairly well proven that female chickens recovering from white diarrhea become bacillus carriers. In such birds these bacteria are found particularly in the ovary. In this

Fig. 60. — The normal ovary of a laying hen. (After Rettger, Kirkpatrick and Jones.)

organ they cause many of the developing ova to become abnormal and undergo degeneration as shown in Fig. 61. Many of the yolks which do not degenerate contain these bacteria. These infect the chick which hatches from such an egg. These chicks then serve to infect others in the incubator or in the brooder. This is undoubtedly one of the reasons that incubator chicks suffer more from this disease than hen hatched chicks. In the incubator there are eggs from a large number of different hens. There is a great probability that one of the mothers may be a bacillus

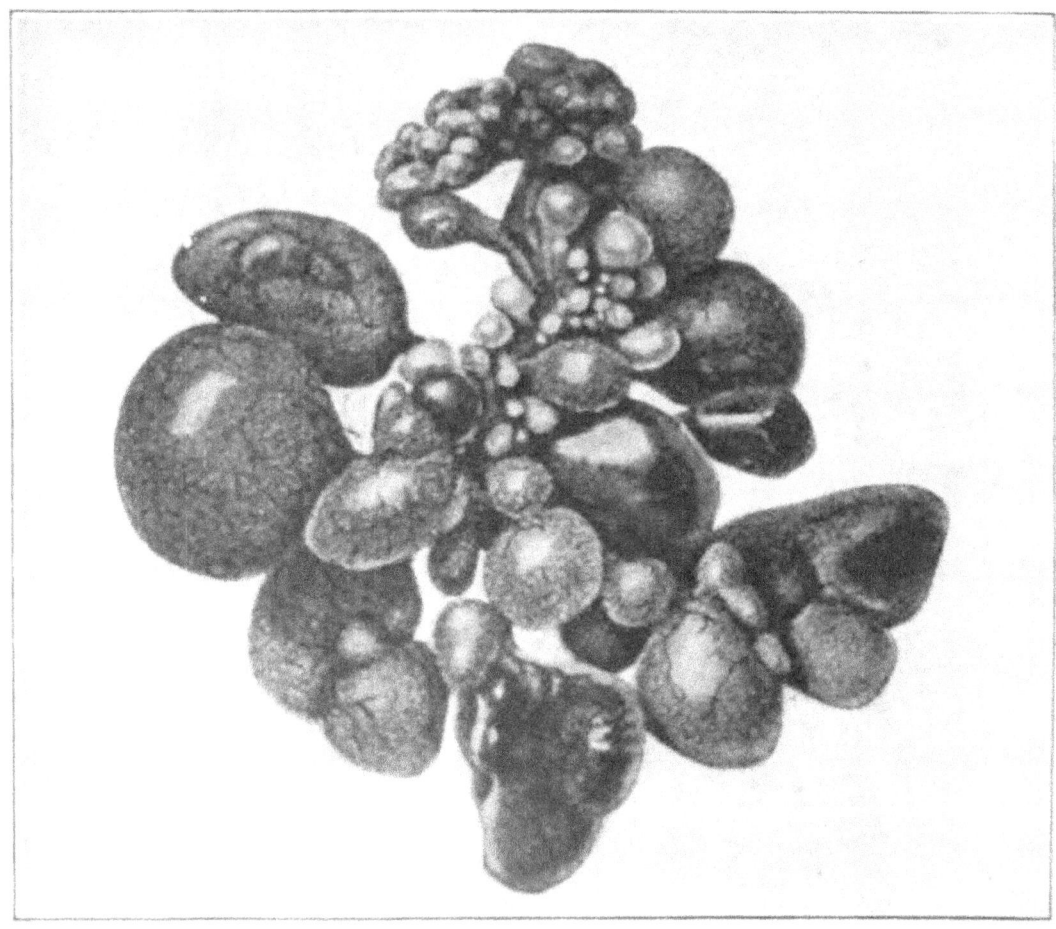

Fig. 61. — An ovary from a hen infected with *B. pullorum*, showing the degenerating and discolored ova. (After Rettger, Kirkpatrick and Jones.)

carrier and one such infected chick may spread the disease to all the others. With a hen there is much less chance that any of the eggs she broods comes from an infected bird. Hence, while some broods will be infected, others will not and one easily gains the impression that hen hatched chicks are less susceptible. This point will be discussed in connection with prevention (p. 299).

The following figure taken from the Storrs Experiment Station Bulletin 68 shows diagrammatically how the infection perpetuates itself from the hen to egg and the chick and from the recovered chick back to the hen again.

Recently Rettger[1] and others have shown that healthy adult fowls may become infected both by contact with infected fowls and through infected litter. This adds

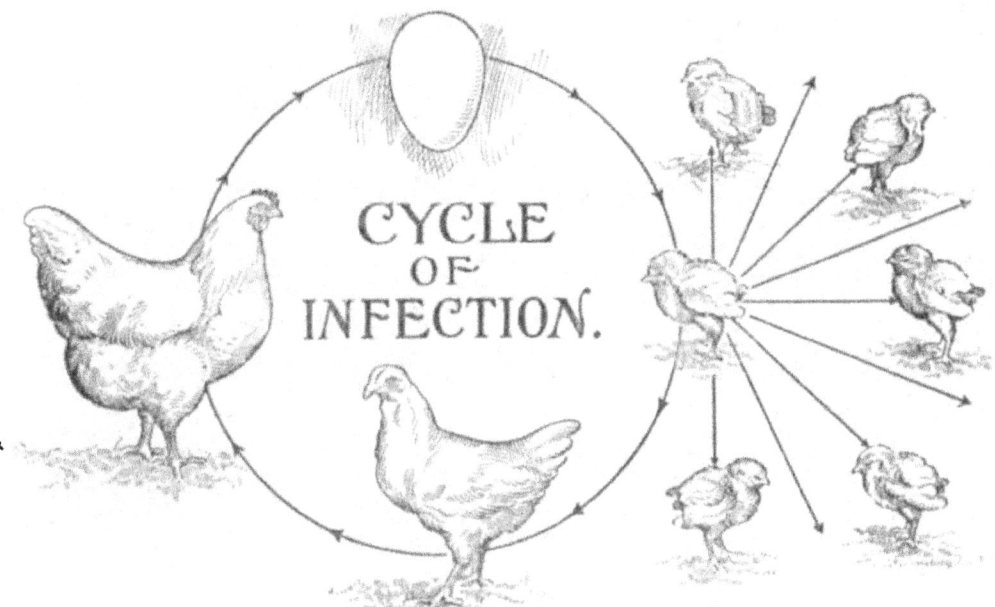

Fig. 62. — Showing how bacillary white diarrhea perpetuates itself in the breeding stock. (After Rettger and Stoneburn.)

another means by which the infection can be spread through the flock.

In this connection it is worth pointing out that Smith and Ten Broeck[2] have found that the bacillus of fowl typhoid shows very many points of resemblance to *B. pullorum*. The only differences found between these bacilli are in respect to their ability to ferment the sugars, dextrose,

[1] Rettger, Kirkpatrick and Jones, *loc. cit.*
[2] Smith, T., and Ten Broeck, C., "A Note on the Relation between *B. pullorum* (Rettger) and the Fowl Typhoid (Moore). *Jour. Med. Research*, Vol. 31, pp. 547–557, 1915.

mannite and maltose. Even these differences appear to be lost if strains of *B. pullorum* are used which have been cultivated in the laboratory for some time. These studies, however, are not extensive enough to justify as yet the assumption that the two organisms are identical.

Diagnosis of White Diarrhea. — The symptoms of white diarrhea are in general the same for the different forms of the disease. They may be briefly stated as follows: The affected chicks appear stupid and remain under the hover or hen much of the time. They isolate themselves from the rest of the flock and appear indifferent to what goes on

FIG. 63. — Ten day White Leghorn chicks showing symptoms of bacillary white diarrhea. (After Rettger and Stoneburn.)

about them. Their feathers become rough and the wings droop (cf. Fig. 63). There is progressive loss of weight. The birds eat little or nothing and appear unable to pick up their food. Their actions in this direction are chiefly mechanical. The characteristic whitish discharge from the vent very soon makes its appearance. The discharged matter may be creamy or sometimes mixed with brown. The discharged matter is more or less sticky or glairy. In many cases it clings to the down in sufficient quantity to plug up the vent. This condition is known as "pasting up behind." This latter condition, however, is not necessarily indicative of white diarrhea.

Many of the chicks chirp or peep constantly or will utter a shrill cry apparently of pain, when attempting to void the excreta. These sounds are often characteristic of the disease.

In many cases the chicks present the appearance of being "short backed" or "big bellied." Woods [1] describes this as follows:

"The weakling is almost always big-bellied, the abdomen protruding to the rear so that it bunches out behind, well out of line with the vent, with the result that the chick looks as if

FIG. 64. — Normal ten day White Leghorn chicks. (After Rettger and Stoneburn.)

the tail piece and backbone has been pushed forward and in just above the vent" (cf. Fig. 63).

In some cases the chicks die with but little warning and show few of the above symptoms. In other cases the sick chick will last a long time showing all the symptoms mentioned.

Post-mortem examination often reveals but few lesions. One of the most striking things is the loss of flesh if the disease has lasted for some time. The alimentary canal is usually nearly empty except for some slimy fluid. The

[1] Woods, P. T., "Reliable Poultry Remedies." Quincy, Ill.

organs are all very pale. The liver may have a few streaks showing congested areas. Some of the unabsorbed yolk may or may not be present. There is considerable variation in its appearance. It is not usually putrid unless the chick has been dead for some time.

The following post-mortem appearances in bacillary white diarrhea are reported by Rettger and Stoneburn:[1]

"Crops — Empty or partially filled with slimy fluid or with food.

"Lungs — Apparently normal. (Tubercles not observed.)

"Liver — Pale, with streaks and patches of red. These apparently slightly congested areas are usually large in size.

"Kidney and Spleen — Apparently normal.

"Intestines — Pale, and for the greater part empty. A small amount of dark grayish or brownish matter frequently present.

"Ceca — With few exceptions but partially filled with a grayish soft material. Only occasionally cheesy or firm contents.

"Unabsorbed Yolk — Usually present varying in size from a pea to a full-sized yolk. The color may vary from yellow to brownish green or nearly black. In consistency there is also much variation. It may appear perfectly normal, distinctly gelatinous, or watery. Frequently it is observed in the character of custard and again more or less dry and firm. Unless the chick has been dead for some time the yolk is usually not found putrid, but merely stale.

"The chick as a whole appears more or less anæmic and emaciated. The muscles of the wings, breast and legs may be almost completely wasted away."

[1] Rettger, L. F., and Stoneburn, F. H., "Bacillary White Diarrhea of Young Chicks." Conn. (Storrs) Agric. Exper. Stat. Bull. 60, pp. 33–57. 1909.

The chief difference between this and coccidiosis appears to be in the contents of the ceca (cf. p. 286).

Undoubtedly the most specific method of diagnosing white diarrhea is by bacteriological examination. Rettger and Stoneburn [1] have perfected methods for the examination of eggs, dead chicks and the ovaries of laying hens. By these methods it is possible for a bacteriologist to determine whether the chicks or the laying stock are infected with *Bacterium pullorum*. Such examination cannot be made by the poultryman. In some states the Experiment Station will undertake such examinations. In others it will be necessary to depend upon private laboratories. The following bacteriological description of *Bacterium pullorum* is taken from Rettger, Kirkpatrick and Jones.[2] It is inserted here for the convenience of bacteriologists who may wish to study this disease.

Description and General Characteristics of Bacterium Pullorum

Morphology, Staining Properties, etc. — The organism is a long, slender bacillus (0.4–$0.5\mu \times 2$–4μ) with slightly rounded ends. It usually occurs single, chains of more than two bacilli being rarely found. It is a non-motile, non-liquefying, non-chromogenic, facultative anaërobe. In its microscopic appearance it resembles the bacillus of typhoid fever. It is stained readily by the ordinary basic aniline dyes. It does not stain by the Gram method; neither does it retain its color when treated with dilute acetic and mineral acids. The organism does not produce spores, or at least they have never been observed.

[1] *Loc. cit.* 1911.

[2] Rettger, L. F., Kirkpatrick, W. L., and Jones, R. E., "Bacillary White Diarrhea of Young Chicks." Conn. (Storrs) Agr. Expt. Stat. Bul. 77, pp. 263–309, 1914.

The thermal death point (moist) is 56 to 57° C. for an exposure of fifteen minutes. The optimum temperature is 35 to 37° C.

Cultural Characters. Agar plates. — Small white colonies make their appearance within twenty-four hours. They increase in size slowly and seldom attain more than one millimeter in diameter, even after three or four days' incubation. Under the microscope they appear yellow and vary in form, being oval, spindle-shaped or round. The surface is usually marked with one or two rosette figures.

Slant Agar. — The ordinary streak growth is quite visible in twenty-four hours, and resembles that of the typhoid bacillus. It spreads little and remains delicate even after prolonged incubation. When, however, the entire surface of the agar is streaked, with a platinum loop, the characteristic cultural appearance of the common pus streptococcus is obtained. The growth is not continuous and compact, but consists of minute, delicate colonies, which may be so small as to require a magnifying lens for detection. This cultural characteristic is of extreme importance in identification work.

Gelatin Plates. — Small white colonies may be seen in forty-eight hours. They remain small for several days, and only under exceptional conditions do they develop into characteristic surface colonies which to a certain extent resemble the grape-leaf colony of *B. typhosus*.

Gelatin Stab. — A delicate growth occurs in forty-eight hours along the whole line of inoculation. In *litmus milk* little or no apparent change occurs within the first forty-eight hours, after which the milk becomes slightly acidified without any signs of coagulation of the casein.

Gas Production in Sugar Bouillon. — Negative results were obtained with maltose, lactose, saccharose, inulin, and dextrin bouillon. Dextrose and mannite were attacked, however, with both acid and gas production. In the dextrose fermentation tubes about 20 per cent of the closed arm is filled with gas, and the mannite tubes average about the same. The gas consists of CO_2 and H in the ratio of 1 : 3. Some of the strains do not produce gas in any of the sugar media, though acid production is quite apparent.

Indol and Nitrate Production. — Neither indol nor nitrate could be detected in Dunham's peptone solution at the end of one week's growth in the incubator.

Quite recently still another method for the diagnosis of this disease has been proposed. Jones [1] points out that the bacteriological examination of eggs is an unsatisfactory method of detecting fowls that are harboring this germ in the ovary. This arises from the fact that with such an infected hen only occasional eggs may contain this organism. Yet such a hen is a source of danger to the flock. Jones then suggested the use of the agglutination test. This test consists in adding a very small quantity of the blood serum of a suspected fowl to a dilute (milky) suspension of the bacteria. If the fowl has the disease the bacteria will clump together and settle to the bottom of the test tube so that the liquid above will appear clear. If the bird is not affected the bacteria will not clump together and the suspension will retain its milky appearance.

This test has now been used by Jones, Gage [2] and Rettger, Kirkpatrick and Jones.[3] In all of these studies it has proved to be an important aid in the recognition of this infection in laying hens.

[1] Jones, F. S., "The Value of the Macroscopic Agglutination Test in Detecting Fowls that are Harboring *Bacterium pullorum*." *Jour. Med. Res.*, Vol. 27, pp. 481–495, 1913.

[2] Gage, G. E., "On the Diagnosis of Infection with *Bacterium pullorum* in the Domestic Fowl." Mass. Agr. Expt. Stat. Bul. 148, pp. 1–20, 1914.

[3] *Loc. cit.* 1914.

This like the bacteriological test cannot be made by the poultryman. It must be done in a well equipped laboratory and under the direction of a competent bacteriologist. The practicability of the test depends upon the value of the fowls.

Treatment. — The treatment of white diarrhea like that of most other poultry diseases consists in prevention rather than cure. Proper care of the breeding birds, proper incubation and proper care and feeding of the chicks will do much to prevent the ravages of this disease. A chick that lacks constitutional vigor, or that is weak from improper incubation or improper brooding falls an easy prey to an infectious organism. On the other hand, a healthy vigorous chick will resist the attacks of such an organism for some time.

However, care in housing and raising the chicks is not sufficient to prevent this disease in a badly infected flock. Undoubtedly, in the light of the researches reviewed in the preceding pages, the best point to attack this disease is the laying hen. Recent work has shown that these infected hens can best be identified by the agglutination test. Where it is possible to apply this test, all reacting birds should be removed from the breeding pens and not allowed to come in contact with the healthy birds. Such a method accompanied by care and cleanliness in raising the chicks will practically eradicate the disease.

There are many poultry plants, however, where it is impossible or impracticable to apply this test. Under such circumstances there is no certain way of identifying the bacillus carriers in the breeding pens. The following method will greatly aid in reducing the mortality from this disease although it will not eradicate it completely.

In the first place every incubator and brooder should be thoroughly disinfected before using and between each hatch.

This can easily be done by spraying with cresol soap (see p. 17) or some other good disinfectant. This will insure that no germs remain from the preceding hatch. Next some kind of wire trays or baskets should be provided which hold from 12 to 15 eggs each. These should be made with a cover and of such shape that they will conveniently fit into an incubator tray. On the eighteenth day the eggs should be placed in these trays and the lids carefully fastened. The ideal method is to have the eggs from each hen in a separate tray but where trap nesting and pedigree breeding are not carried out this is impracticable. The chicks should be allowed to hatch in these trays and to remain in them until they are 48 hours old. By this time they have passed the most critical stage and they may then be put together in the brooder. By thus isolating the chicks in small groups only a few of these groups will usually prove to be infected. It has already been pointed out that one infected chick will spread the disease to an entire incubator or brooder if allowed free range among its neighbors at the critical period. This method has been used with marked success on a number of large poultry plants.

During the last few years there have appeared many articles in the poultry press regarding the use of sour milk as a cure or preventive of white diarrhea. The Storrs Experiment Station has carried out careful experiments in this connection extending over several years. Their results are summed up in a recent bulletin.[1]

"Sour milk feeding has a most beneficial influence on the growth of chicks and in lessening mortality from all causes. As an important agent in the prevention and suppression of white diarrhea its value is somewhat doubtful, and further investigation is necessary before unqualified statements

---

[1] Rettger, Kirkpatrick and Jones, *loc. cit.*, 1914.

can be made. Milk which is soured by the Bulgaricus bacillus of Metchnikoff possesses no distinct advantages over naturally soured milk; on the other hand, it has several disadvantages. Its method of preparation involves considerable time and care, and it is not relished by chicks to the same extent as naturally soured milk."

Other remedies have been proposed for this disease but most of them appear to have but little value. Kaupp[1] recommends a sulphocarbolate treatment as follows: To each gallon of drinking water he dissolves one 30-grain tablet sulphocarbolate compound of zinc, sodium and calcium, 6 grains of bichloride of mercury and 3 grains of citric acid. With this mixture he reports good results. Horton[2] at the Oregon Experiment Station used this treatment on 50 infected chicks of which only seven survived. He says: "From the manner in which the chicks died off and from the general appearance of the seven that remained alive it seems evident that sulphocarbolates in the treatment of white diarrhea (bacillary form) have very little if any efficiency."

[1] Kaupp, B. F., "Some Poultry Diseases." Colo. Agr. Expt. Stat. Bul. 185, 1912.

[2] Horton, G. D., "Sulphocarbolates in the Treatment of White Diarrhea (bacillary form) of Young Chicks." *Amer. Vet. Rev.*, Vol. 46, pp. 321–322, 1914.

CHAPTER XIX

OTHER DISEASES OF CHICKENS

Leg Weakness

THE term "leg weakness" is sometimes used by poultrymen to indicate the lameness due to rheumatism in adult birds. Regarding this form of the disease see page 201. The more usual use of the term "leg weakness" is to denote a disease or ailment which is found in growing chicks from one month to six months of age. It is said to be more common among cockerels than pullets and is more frequent in the heavier than the lighter breeds. The chief cause of the trouble seems to be that in birds growing rapidly and fed heavily the weight sometimes increases faster than the strength. This results in a weak kneed, wobbling bird. The disease is sometimes ascribed to other causes such as overcrowding, close, unventilated quarters, overheating, etc. Salmon says, "It may develop in young chickens kept in brooders in which the heat is not properly distributed or where there is too much bottom heat, also in those which are kept constantly upon wooden floors." Regarding these cases Robinson says, "Where such conditions are present the leg weakness is more likely to be an accompaniment of diseases which plainly show other symptoms."

Diagnosis. — The symptoms are indicated in the name of the disease. It first appears as an unsteadiness in the walk. This may gradually become worse until the bird is unable to stand alone and is constantly tumbling over. The birds are found sitting while eating and are inclined to walk very little. When the trouble first appears there is little else wrong with the bird. The eye and comb are bright and healthy, the appetite is good. Later, however, the bird being weaker than the others gets less grain and becomes thin, feathers out poorly and is a distressed object. It is said that rheumatism can be distinguished from leg weakness by the swelling of the joints in the former disease.

Treatment. — This consists chiefly, of course, in removing the cause. Since the most common cause is the overfeeding with fat producing foods, the amount of these should be reduced. The weak birds should be removed to a pen by themselves. Substitute bran, wheat and oatmeal for the corn and cornmeal. Give skim milk, if possible, instead of water. *Feed plenty of green food.* This is one of the most important measures. Sanborn recommends rubbing the legs with tincture of arnica and adding ½ teaspoonful of tincture of nux vomica to each quart of drinking water.

Aspergillosis or Pneumomycosis

This disease, which is discussed on page 173, not only occurs in hens but it is also a very common and fatal disease in young chicks. It often occurs with white diarrhea and the double disease was for a long time considered as one. Poultrymen designated the cases in which the lesions occurred in the lungs as "lungers." Investigation has shown that there are two diseases which may occur separately or together.

Diagnosis. — This disease is characterized by a dumpish sleepy condition of the chick. The wings are pendulant. Breathing is rapid and sometimes accompanied by snoring sounds. A whitish diarrhea is present. A differential diagnosis between this and the coccidial white diarrhea is only possible by an examination of the dead birds. In aspergillosis, yellowish tubercles which closely resemble those of tuberculosis occur in the lungs and in the walls of the air sacs and often also in the intestines, mesentery, liver and other organs. In very acute cases the lungs are simply inflamed, death occurring before the formation of the tubercles. The mycelium and spores of the fungus may be found by microscopic examination of the tubercles and this fungus may be obtained by inoculating cultures from these tubercles.

Etiology. — The disease is caused by the spores of an *Aspergillus*, usually *A. fumigatus*, Fig. 32. This is a very common fungus and the spores are widely distributed in nature. The spores are often found on the food or on the litter and are inhaled or taken in with the food. Incubator chickens are often infected from the incubators and brooders and hen hatched chickens from the straw or chaff in the nests. Sometimes the chicks get the disease from chick food not properly cared for. It is possible that this disease as well as the coccidial and bacillary white diarrhea is sometimes carried in the egg. The spores and mycelium are often found in the digestive tract of hens and it is not unlikely that they may work up the oviduct from the cloaca and infect an egg before it gets its shell.

Treatment. — The treatment of diseased chicks is useless. When they are infected the spores develop on the membranes and new spores are formed which spread the infection throughout the respiratory system and also to the other organs. The only effective treatment is prevention. Keeping the flock under good hygienic conditions with clean food, litter and nesting material reduces the chance of infection and keeps the chicks in a vigorous condition in which they are able to resist the disease. The dead chicks should be burned or buried.

Prognosis. — The disease is fatal so far as known.

Emphysema

This name is applied to a disease of young chicks in which the skin puffs out in the sides of the neck near its juncture with the body. The size of the puff varies somewhat. In mild cases it is about the size of a hickory nut. Sometimes there is one puff, sometimes several.

According to Vale the trouble generally occurs in growing chicks which have been confined in close quarters. It is often associated with some lung trouble. It seems to be due to obstruction of the air passages and the rupture of some of the air sacs. The air thus escapes into the tissues beneath the skin. While not common this disease does occur in Maine. Some cases were reported to the Station while this work was in preparation.

The treatment suggested by Vale is to puncture the skin with a needle and to give 2 grains nitrate of iron to each wine glassful of drinking water.

Gapes

Gapes is a disease which attacks domestic poultry and many species of wild birds. In fowls it is more frequently observed in young chicks. It occurs also in adult fowls but rarely causes enough inconvenience to attract attention. The disease is due to the presence of minute parasitic worms in the air passages.

Diagnosis. — The characteristic symptoms of this trouble are frequent gaping, sneezing, coughing with discharge of mucus. The affected birds appear weak and dumpish with drooping wings. When badly affected the bird stands or sits with its eyes closed, wings drooped, mouth open and at frequent intervals gasps as if suffocating.

The correctness of a diagnosis for gapes should be tested by

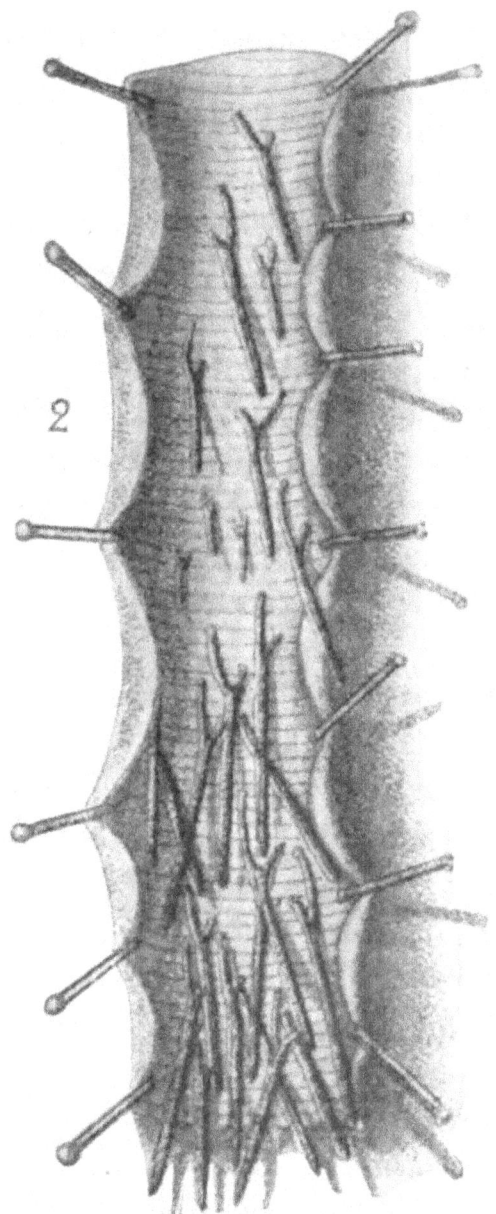
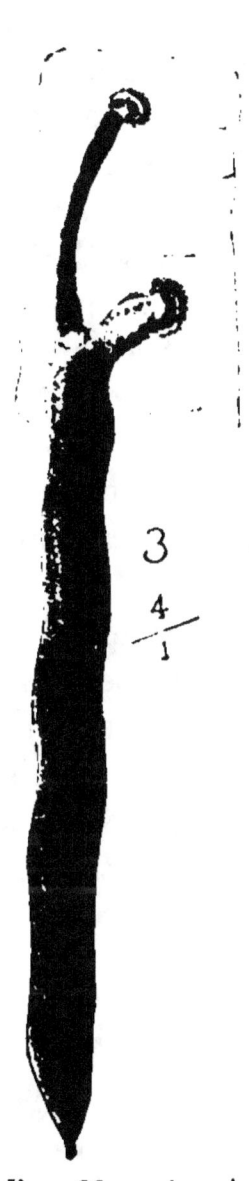

Fig. 65. — Trachea (windpipe) of a pheasant showing gape worms (*Syngamus trachealis*) attached to the mucous membrane. (After Megnin.)

Fig. 66. — A pair of *Syngamus trachealis*, attached. (After Megnin.)

determining whether or not the worms are present in the trachea. When chicks are dying from a disease supposed to be gapes the trachea of a dead bird may be examined. If the trouble is gapes the worms will be found attached in pairs to the mucous membrane of the trachea.

The two sexes are joined together in such a way that a pair looks like a double headed worm. The female is about $\frac{1}{2}$ inch long and the male about $\frac{1}{5}$ inch. The worms are pale in color when empty but when they have been feeding they are red with the blood of the chick. The presence of the worms in the trachea of a living chick may be demonstrated by passing a gape worm extractor (a loop of horse hair or fine wire or a feather with the vane removed except at the tip) carefully down the trachea for some distance turning it around to loosen the worms and drawing it out. If the worms are present some will be removed with the extractor.

The presence of the worms causes an irritation and inflammation of the membrane and stimulates the secretion of mucus. Some of the accumulation of worms and mucus is expelled by coughing. Sometimes part of it is swallowed and expelled with the feces. The loosened material may be drawn into the deeper air passages during inspiration. Death may occur from suffocation due to the obstruction of the air passages with worms and mucus, or weak individuals may die from loss of blood.

Etiology. — The only cause of the disease is the nematode or thread worm *Syngamus trachealis* Siebold, called the gape worm, red worm, or forked worm (see Figs. 65 to 67). These parasites obtain their nourishment by sucking the blood from the mucous membrane of the trachea. They are attached in pairs to the membrane by their sucker-like mouths. Besides bringing about a considerable loss of blood the worms cause irritation and inflammation of the membrane and a copious secretion of mucus. The two sexes are so closely attached to each other that they cannot be separated with tearing. The body of an adult female is swollen with

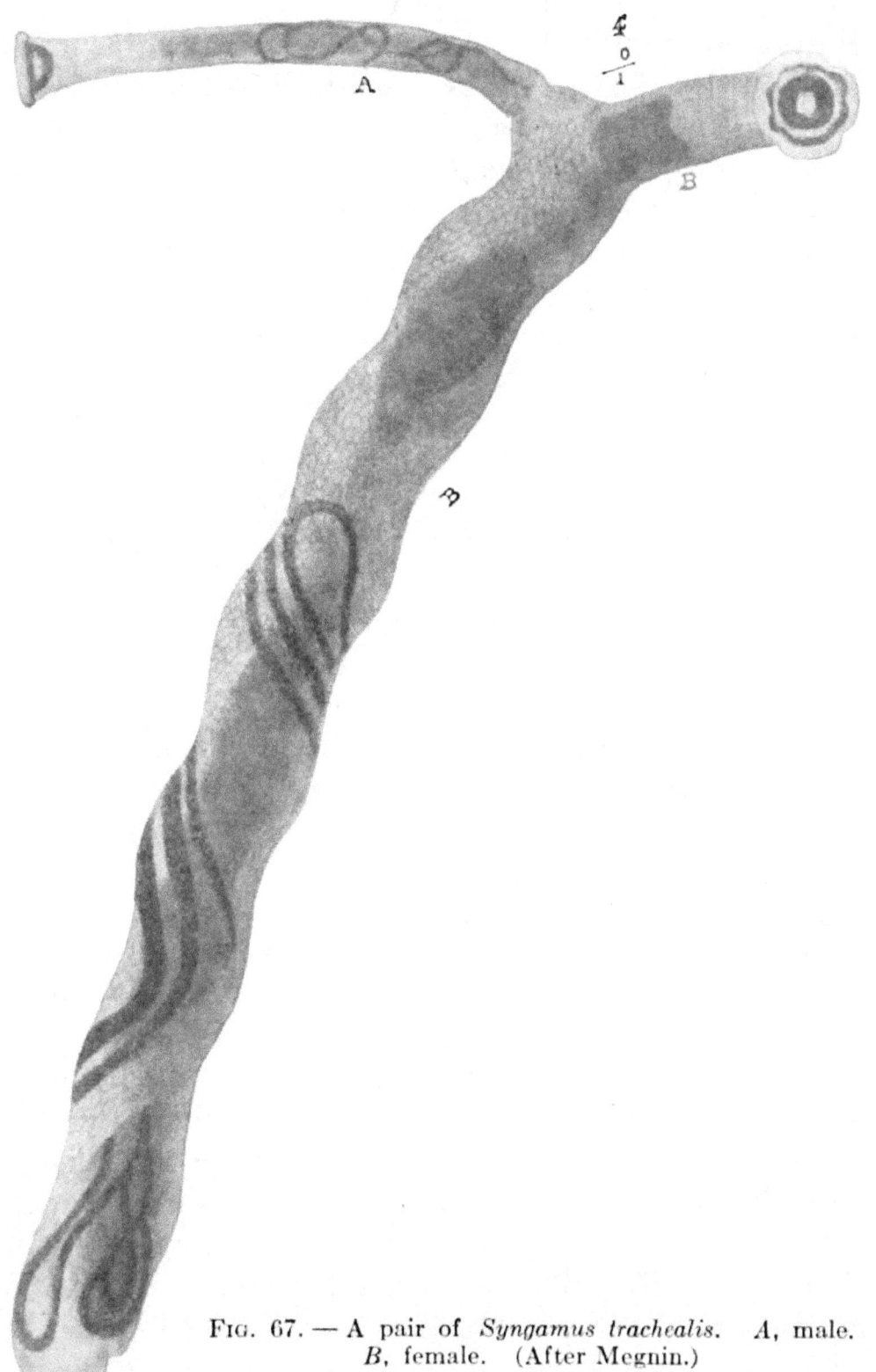

Fig. 67. — A pair of *Syngamus trachealis*. A, male. B, female. (After Megnin.)

thousands of eggs and occasionally contains some embryos. The eggs are not laid but escape when the body of the female is ruptured. This may take place with the decomposition of the worm or the body may be torn by the coughing of the bird. The eggs may develop and grow to adult worms within the trachea of the same bird. The worms, eggs and embryos are often coughed up. Sometimes they are swallowed and then some of the eggs and embryos may be passed with the feces. The worms coughed up are eagerly eaten by the same or other birds and the ova and embryos are often taken with contaminated food and drink. Developing embryos have been found in earth worms living in infected poultry yards, and these will cause gapes if fed to chicks.

The eggs and embryos need only warmth and moisture to develop. Eggs may develop in the digestive organs. It is not known how the embryos reach the trachea from the digestive organs. A large number of those eaten never reach the trachea but are either digested or voided with the feces. Salmon says: "Although there are some thousands of eggs in the adult worms, 10 to 15 worms have been fed to a single chicken, and, as a result, not over 4 or 5 embryos would reach and develop in the trachea." According to Theobald, Ehler found copulated worms where several of the females were full of eggs 10 days after feeding ova to healthy chicks. Wet clay soils are especially favorable to the gape worms, and they thrive best in warm, wet weather.

Treatment. — In eradicating the disease it is important to isolate all affected birds so that the worms and ova coughed up or voided with the excrement may not be eaten by the other chicks or contaminate the food, drink, and the ground of the runs. Burn the bodies or at least the heads and necks of all dead birds. The feed troughs and water dishes should be scalded and the houses and coops disinfected. Use potassium permanganate in the drinking water. If possible provide fresh runs on which there has been no poultry for several years.

The following methods have been recommended for disinfecting the ground. It is doubtful if these are economically advisable.

Treating the ground with air slaked lime and spading.

Sprinkling with one of the following solutions:

1 per cent or 2 per cent sulphuric acid.

2 ounces of copperas dissolved in a pail of water.

½ ounce of crystals of potassium permanganate to a barrel of water.

The lime or acid treatments are most often recommended. The infected birds should be kept in houses easily cleaned and disinfected and this should be done frequently to prevent reinfection of the recovering birds. Theobald advises an addition of 3 drams of salicylate of soda to each quart of drinking water to destroy eggs and embryos that may contaminate it.

The individual surgical method may be profitably practiced in some cases. It seems to be the only sure method yet advised of ridding an infested bird of the parasites. Wright[1] gives the following description of the method:

"The old-fashioned cure was to strip a small quill-feather, all but a small tuft at the point, and (moistening it in turpentine or not) introduce it into the trachea, turn it round, and withdraw it with the worms. This is effectual, but requires care to prevent lacerating the windpipe or causing suffocation. In this way 30 worms have been successfully extracted from one chicken. A very much better method is to take two *straight* hairs from a horse's tail, laid together, tie a knot on the end of the pair, and cut off the ends close to the knot. This is passed straight (*i.e.*, without twisting) down the windpipe as far as it will go without

[1] Wright, L., "The New Book of Poultry." London, 1905.

bending, then twisted between the finger and thumb and drawn out. A trial or two may miss, but usually 5 or 6 attempts will bring up 4 or 5 worms, and the hairs inserted in this way, without twisting, do not seem to hurt the chicks, and are used with the greatest facility. The bringing up of even from 4 to 10 worms, and the failure of more to come after a blank trial or two, may usually be reckoned as a cure."

Wire gape worm extractors may be bought from dealers in poultry supplies, or one can make one for himself by taking No. 30 wire, forming a loop at one end just big enough to go easily down the trachea, and then twisting together the ends of the wire to form a long handle. Worms removed should be burned.

It is reported (*Jour. Bd. of Agric.*, London, Vol. 13, p. 368, 1906) that gapes may be successfully treated by the fumes of carbolic acid. The method given is to place the chicks in a basket over a pail containing carbolic acid. A hot brick is placed in the pail for the purpose of volatilizing the acid.

Prognosis. — This disease is often fatal in young chicks from one to four weeks old, especially in small, weak birds. In young chicks and in most adult fowls it often causes little inconvenience. These fowls, however, are constant sources of infection. The removal of the worms from the trachea if skillfully done so that the delicate membrane is not injured usually effects a cure but this individual treatment requires considerable time and the value of the chicks must determine whether or not it is economically profitable.

Crooked Breast Bone

The normal breast bone of a fowl is shaped like a boat with a deep keel. This keel is a thin plate of bone which furnishes a place for the attachment of the large flying muscles. In a fully mature bird the breast bone is completely ossified but at hatching it is almost entirely cartilaginous.

Normally the keel of the bone is straight and perpendicular to the basal portion of the bone. In many individuals, however, it is bent to one side or first to one side and then the other forming an S-shaped curve. Also it may be inclined at the base at an abnormal angle.

Poultrymen usually attribute these abnormalities to the fact that the birds go to roost too young. They believe that they are caused by the pressure of the hard roost on the soft bone. About fifty years ago Rottiger [1] called attention to the fact that crooked breast bones occur in chicks showing retarded growth before they have ever roosted and also in wild birds kept in captivity and prevented from early roosting. He also found these malformations associated with diseased conditions, especially catarrh. He believed that they were due not to external pressure but to a lack of bone forming elements in the diet or to a derangement of the digestive apparatus which prevented the proper assimilation of these elements.

Our own experience is in accord with this view. Crooked breast bones often occur in cases of malnutrition or disease without reference to whether the birds have or have not roosted. In utility stock the crooking of the bone in itself may not be of great importance, though it lowers the sale value of a bird greatly. Some good layers have crooked breast bones. However, when large numbers having this malformation occur in the flock there is something wrong with the feeding or care of the chicks and such mistakes should be looked into and corrected.

[1] Rottiger, *The Poultry World*, Vol. 5, p. 298. (Translated from *Ztschr. f. Geglügel u. Singvogelzucht* by W. G. Todd), 1876.

CHAPTER XX

TUMORS

NEW tissue growths or neoplasms (tumors) are by no means uncommon in domestic fowls. It has been the routine practice at the Maine Agricultural Experiment Station for several years to autopsy all birds that are killed for material or data and all birds that die from natural causes. The archives of the laboratory now contain about nine hundred autopsy records sufficiently complete to determine whether or not the birds had tumors and in what organs the tumors were located. These records show that 8.98 per cent of all the birds autopsied had tumors. That is, there were about 90 cases of tumors per thousand birds. The genital organs, at least in the females,[1] were most often affected. In fact 37 per cent of all the tumors found were in the ovary and 19 per cent in the oviduct. Twenty-two per cent were found in the peritoneum (some of these were attached to the walls of the abdominal cavity and some were in the mesenteries). Tumors have also been found in the intestine walls, kidney, gizzard, liver, spleen, pancreas, heart and breast bone.

Some of these tumors occurred in fowls killed for dissection or data and in apparently normal health. The tumors were the probable cause of death in less than half the cases of birds with tumors which died from natural causes.

The tumor was usually confined to one organ, but there

[1] Very few males were autopsied. One of the three tumors found in males was located in one of the testes.

were fifteen cases in which tumors of similar nature occurred in two or more organs. In these cases the tumor had probably undergone metastasis.

The age of the bird is also related to frequency of the occurrence of tumors. Tumors being much more frequent in old than in young birds. Only 7.37 per cent of the birds under two and one-fourth years of age had tumors, while 19.17 per cent of those older than this were affected.

A further proof of the common occurrence of new tissue growths in fowls is the fact that in the course of ten months Rous, Murphy, and Tytler [1] obtained without difficulty about thirty spontaneous tumors in living fowls.

Fowl tumors are apparently in every way analogous to tumors in human beings. They are masses of new tissue (neoplasms) which persist and grow independently of the surrounding structures. These growths are of no physiological use to the host but they are often harmless. Tumors may be classified as *malignant* or *benign*. Benign tumors do not penetrate into the surrounding tissues but push them aside. They are usually encapsuled. On the other hand there are a number of tumors which are *malignant* and tend to infiltrate the tissues. Many of these produce growth in adjacent organs and even in distantly removed parts of the body. Usually they affect the general health and when removed tend to recur.

On examining four thousand hens brought to a hotel, Ehrenreich [2] found seven malignant tumors. All of these were found in one thousand hens more than one year old.

---

[1] Rous, P., Murphy, J. B., and Tytler, W. H., "A Filterable Agent the Cause of a Second Chicken Tumor, an Osteochondrosarcoma." *Jour. Amer. Med. Assoc.*, Vol. 59, pp. 1793–1799, 1912.

[2] Ehrenreich, M., and Michaelis, L., "Ueber Tumoren bei Hühnern." *Zeitschr. f. Krebsforch*, p. 586, 1906.

Ehrenreich, M., "Weitere Mitteilungen über das Vorkommen malignes Tumoren bei Hühnern." *Med. Klin.*, Berlin, III, 614, 1907.

In the three thousand hens under one year old no malignant tumors were found.

There are many theories regarding the origin of tumors. One which has been quite generally accepted being that unused embryonic cells may remain collected in certain spots and that they may later be sufficiently stimulated to grow independently. In accord with this theory is the fact that the histological structure of a tumor resembles one or another of the general classes of body tissues, and in fact usually copies more or less closely the structure of the organ in which the primary tumor arises. However, when growths arise secondarily in other organs, that is, when the tumor has undergone metastasis, the secondary tumors are similar in structure to the primary tumor. Further Rous and his colleagues [1] have found several distinct chicken tumors which may be produced in healthy fowls by the injection of a cell free filtered extract of the tumor. In the case of each of these tumors the neoplasm produced in the inoculated fowl always resembles the tumor from which the extract was made. That is, the type of tumor is determined by the individuality of the causal agent and not alone by the potentialities of the stimulated cells. That is, the cells are not simply stimulated to grow, but they are stimulated to grow in a specific way.

According to the general type of body tissue they resemble, tumors may be classified as follows:

[1] Rous, P., "A Transmissible Avian Neoplasm (Sarcoma of the common fowl)." *Jour. of Exper. Med.*, Vol. XII, pp. 696–705.

Rous, P., Murphy, J. B., and Tytler, H. W., *loc. cit.*

Rous, P., and Lange, Linda B., "The Characters of a Third Transplantable Chicken Tumor Due to a Filterable Cause. A Sarcoma of Intracanalicular Pattern." *Jour. Exper. Med.*, Vol. XVIII, pp. 651–664, 1913.

Rous, P., and Murphy, J. B., "On the Causation by Filterable Agents of Three Distinct Chicken Tumors." *Jour. Exper. Med.*, Vol. XIX, pp. 52–68, 1914.

I. Connective tissue type.
II. Epithelial tissue type.
III. Muscle tissue type.
IV. Nervous tissue type.

Each of these groups may be subdivided according to the more specific resemblance of the histology of the tumors to a particular kind of body tissue. At least six different types of the first group have been described in the domestic fowl.

I. Tumors of the connective tissue type.

1. Sarcoma (made up of embryonic connective tissue composed of closely packed cells embedded in a fibrillar or homogeneous substance).

2. Myxoma (mucous tissue, a soft translucent growth, made up of variously shaped cells of connective tissue and capillary blood vessels incased in a jelly-like matrix.

3. Fibroma (composed mainly of fibrous or fully developed connective tissue). The tumors of this group are benign.

4. Chondroma (a cartilaginous growth).

5. Osteoma (bony tissue usually but not always developing on bone).

6. Lymphoma (made up of lymphoid tissue).

According to Tyzzer and Ordway [1] tumors of the sixth group (lymphomata) occur more frequently than any other type. They suggest that this is probably due to two facts. (1) This type of tumor apparently develops in younger birds than other types, and (2) birds kept for egg production are usually killed at the end of their first year. They have described in detail seven different tumors of this group. In summing up their results they say: "In some cases lymphoma occurs as a local primary growth either with or without metastasis. In other cases the tumor tissue is so uniformly disseminated throughout certain organs that it is impossible

---

[1] Tyzzer, E. E., and Ordway, T. "Tumors in the Common Fowl." *Jour. Med. Research*, Vol. 21, pp. 459–477, 1909.

to determine the point of origin. Certain lymphomata are more or less alveolar in structure; others grow diffusely through the tissues. The tumor may be confined to the fixed tissues so that it is essentially extravascular, or the tumor cells may also occur in the circulatory blood constituting a lymphatic leukemia."

Probably the second most frequent type of connective

Fig. 68. — Sarcoma Chicken Tumor No. I, Second Generation. (After Rous.)

tissue tumor is sarcoma. This may occur as simple sarcoma or it may occur in combination with one or more of the other types. That is, it is a generalized and simple tissue which may represent the complete adult stage of a tumor, as Rous Chicken Tumor No. I, and Chicken Tumor No. XVIII.

However, since it is a simple generalized tissue, it may also represent an early stage in the development of one of the other forms of connective tissue tumors. In these cases it remains as a matrix in which the further developed tissues

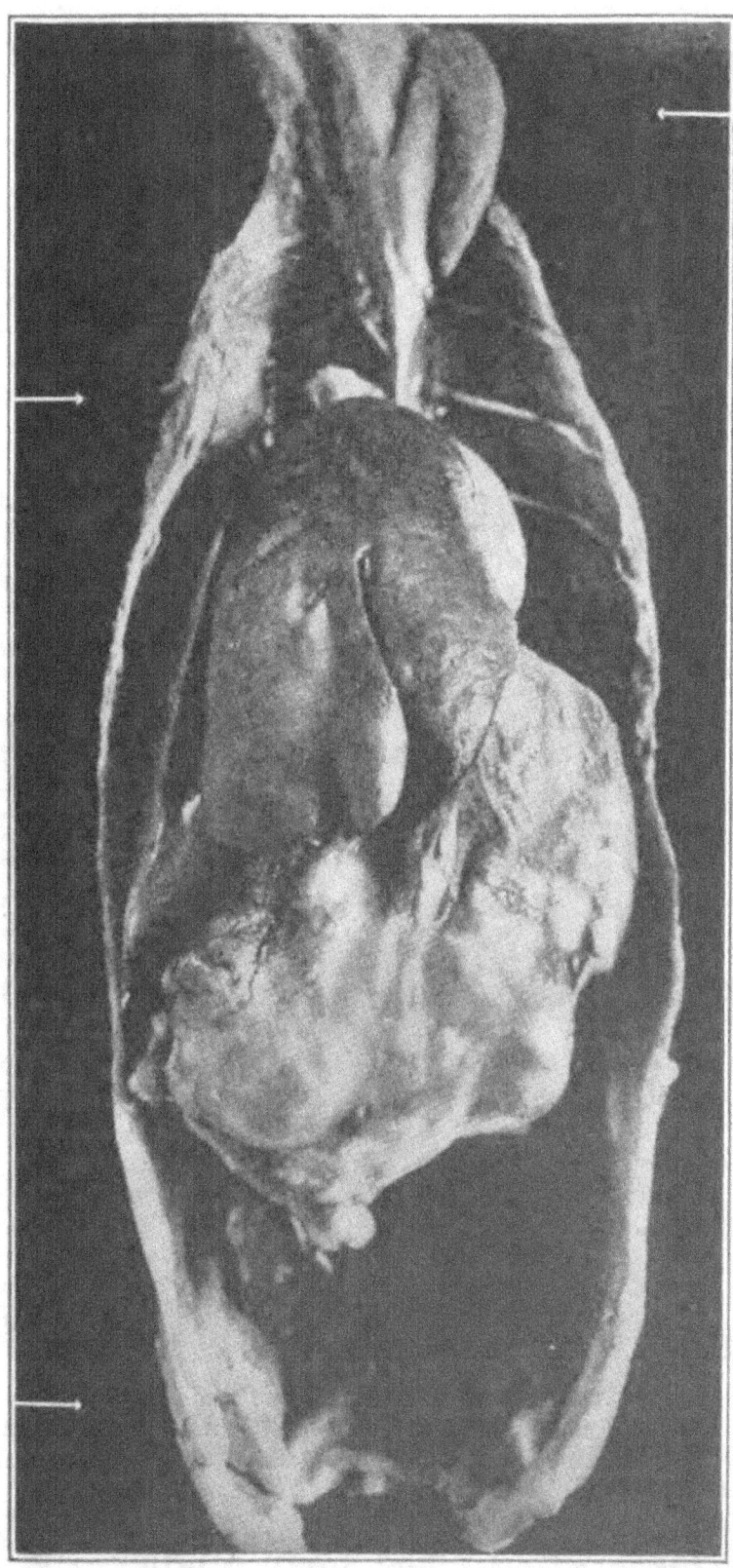

Fig. 69. — Chicken tumor XVIII in the gizzard of the original fowl. Three metastases are visible in the skeletal muscles, namely one in the neck, another in the thoracic wall above the right lobe of the liver and the third on the inner surface of the pelvis. All are indicated by arrows. (After Rous and Lange.)

are embedded. Tyzzer and Ordway[1] have described a myxosarcoma in which masses of mucus were embedded in a

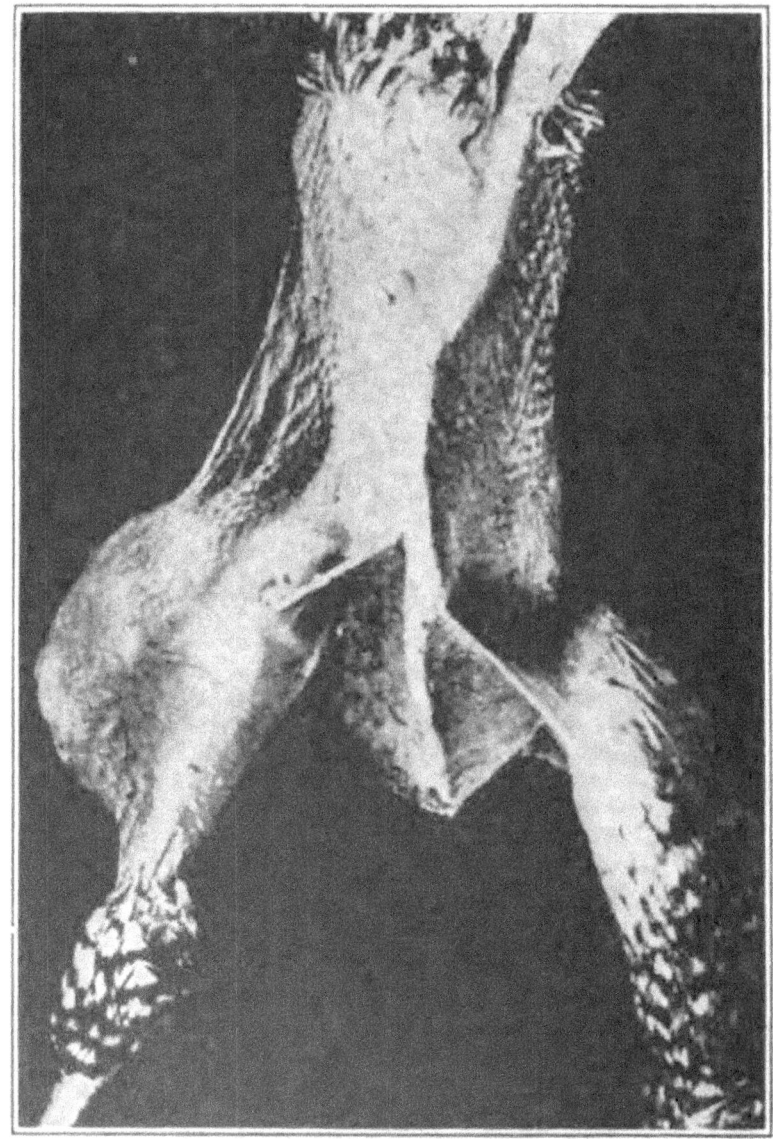

Fig. 70. — Large osteo-chondrosarcoma produced by intramuscular injection of 4 cc. of the Berkfeld filtrate of an extract of chicken tumor VII. The fowl was killed when comatose eighty-seven days after the injection. Its emaciation should be noted. (After Rous and Murphy.)

connective tissue matrix and Rous, Murphy, and Tytler[2] found a transmissible Osteo-chondrosarcoma.

[1] *Loc. cit.* [2] *Loc. cit.*

When the filtered extracts of this tumor were injected into susceptible individuals a rapid growth of cartilage forming connective tissue elements took place. These growths soon became cartilaginous and finally bony.

The connective tissue tumors (especially the sarcomas)

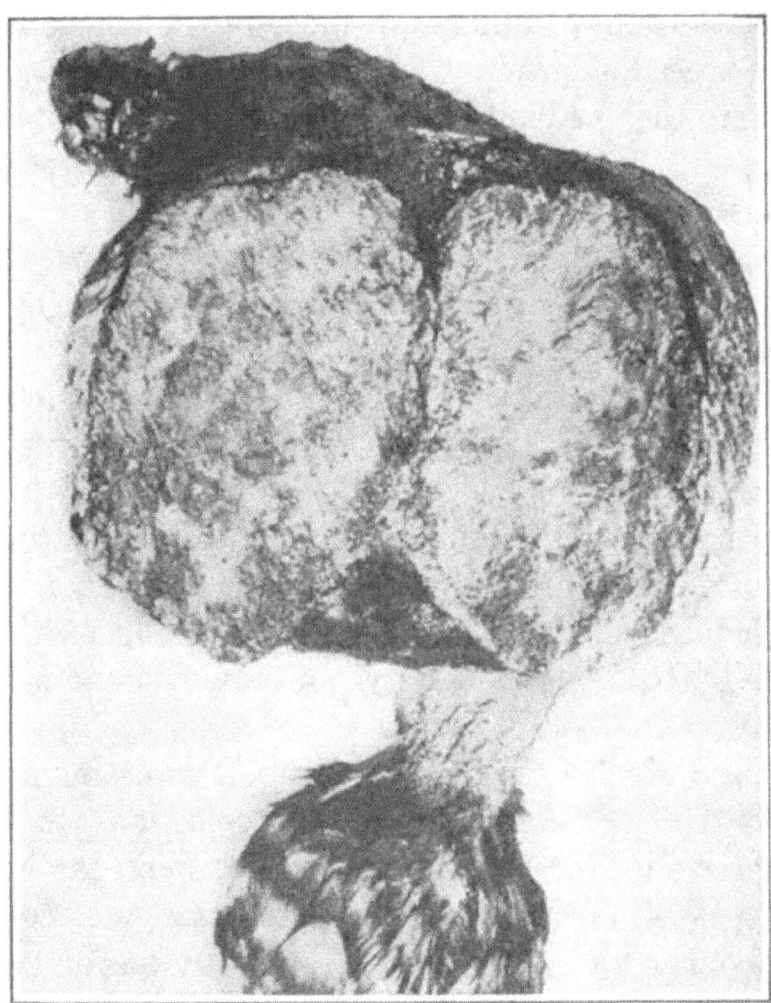

Fig. 71. — The growth shown in the preceding photograph after it had been sawed open. Scattered amid the smooth whitish cartilage is much bone with red marrow. (After Rous and Murphy.)

are of frequent occurrence in mammals, and a large number of the avian tumors which have been described belong to this group. These tumors are sometimes benign but are often very malignant and appear in many organs of the body and frequently cause emaciation and death. All three of the chicken tumors which Rous and his colleagues have found capable of transference to other individuals belong to this group. Tumors of this group are usually covered with a hard, tough, fibrous tissue.

The next group of tumors are those of the epithelial type. As in the preceding group, these tumors are further classified according to the particular type of body tissue they resemble. Two types of these epithelial tumors have been described in the domestic fowl.

1. Adenoma (tumors with a gland-like structure).

2. Carcinoma (cancer — epithelial cells developing in epithelial tissue).

Pickens [1] has described an adenoma of the bile ducts in the domestic fowl (Fig. 72). This tumor was evidently of a highly malignant type as it had been transferred from the liver to the proventriculus, gizzard, spleen, intestines and peritoneum. The growths were small lobulated masses, the larger ones contained small cysts (sacs filled with serous liquid). The abdominal cavity contained about a pint of this liquid.

Pick [2] and Koch [3] have each described a carcinoma found in the mouth of a fowl and Ehrenreich [4] considered that five of the seven malignant tumors he studied were carcinomata. Three types of carcinoma have been described: squamous celled, alveolar and granular cancer. Tumors of this class are malignant.

[1] Pickens, E. M., "A Cysto-adenoma in a Fowl." Rept. of New York State Veterinary College, pp. 261–268, 1913–1914.

[2] Pick, L., "Zur Frage von Vorkommen des Carcinoms bei Vögeln: Grossen Plattenepithelkrebs des Mundhohlenbodens dei einem Hühn." *Berliner klin. Wochenschr.*, 29, 669, 1903.

[3] Koch, M., "Geschwülste bei Tieren." *Verhandl. der deutsch. Gesellsch. f. Path.*, 136, 1904   [4] *Loc. cit.*

Tumors of the third group are formed of muscle-like tissue. In some mammalian tumors (rhabelomyoma) the muscle cells are striated. In others (leiomyoma) they are unstriated. So far as we know no tumors of the first group have been described in fowls, but Tyzzer and Ordway [1] have described a tumor in the mesentery of a domestic fowl which was composed of typical smooth muscle fibers.

At present we know of no description of tumors of the fourth or nervous tissue group in the domestic fowl. Such tumors, however, very likely exist.

We have seen that there are many different kinds of new tissue growths or neoplasms in the domestic fowl. Some of these are benign and some malignant. We are entirely ignorant of the cause of these growths when they spontaneously occur in fowls. Only three out of thirty different tumors tested by Rous, Murphy and Tytler [2] can be reproduced in another individual. None of these tumors has been transmitted to healthy fowls kept with those which have developed it. The work of Rous and others has, however, shown that particular neoplasms have specific causative agents which in some cases can be separated from the neoplasm. In the cases studied this was apparently a living virus, although if organisms are present they are ultramicroscopic and able to pass through a Berkfeld filter impermeable to *Bacillus fluorescens liquefaciens*.

The work of Rous has shown that many birds possess a perfect natural immunity to these tumors and that in general young, vigorous birds are most susceptible. Funk[3] has shown that birds stunted by a deficient diet are less susceptible than

---

[1] *Loc. cit.*

[2] *Loc. cit.*

[3] Funk, Casimir., "Studies on Growth: The Influence of Diet on Growth, Normal and Malignant." *Veterinary Journal*, Vol. 21, N. S., pp. 126–132, 1914. (Reprinted from *Lancet*.)

normally developed birds and that the growth of the tumor (Rous sarcoma) is slower in the underdeveloped birds.

It is a well known fact in human pathology that physiological activity stimulates the growth of some tumors, while physiological decline stimulates others. This is undoubtedly true also of fowl tumors.

Non-malignant tumors located on external parts may be

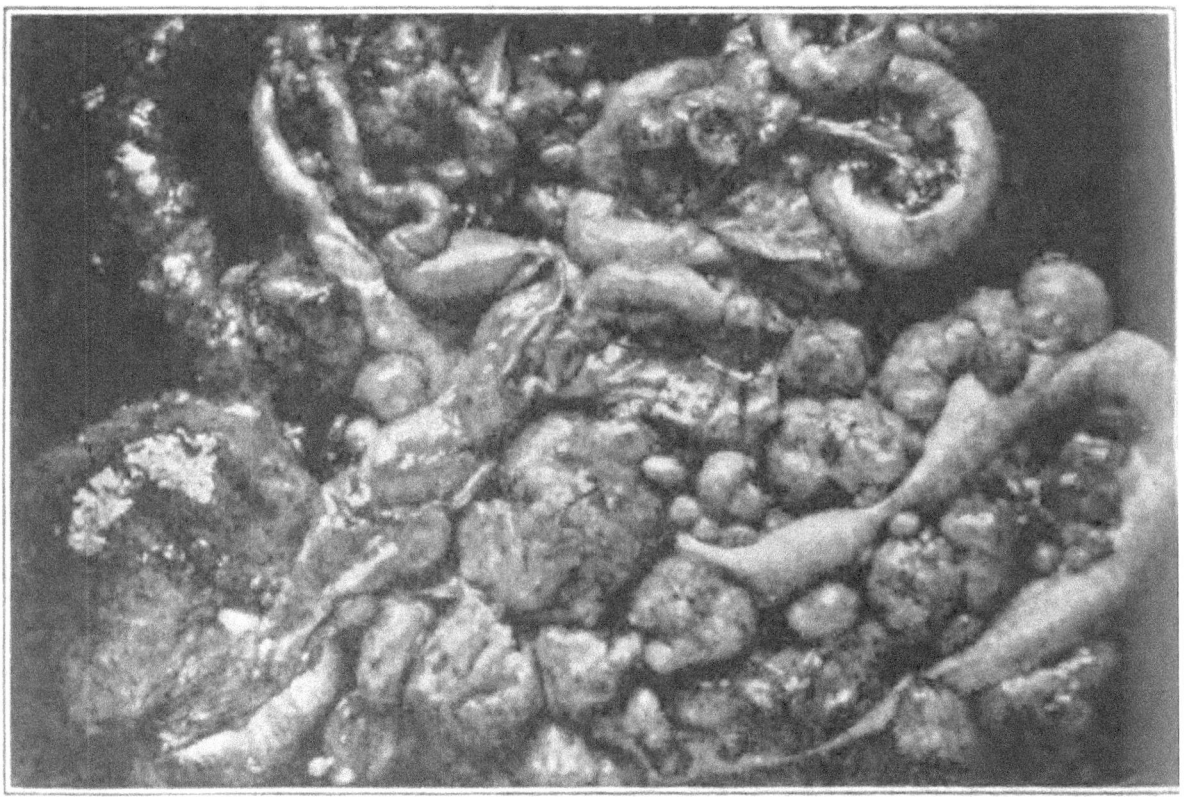

Fig. 72. — Cysto-adenoma on the serosa of the intestine. (After Pickens.)

removed or they may be left alone. They will not ordinarily affect the health of the bird. Even malignant tumors are apparently not contagious.

The ultimate nature of the causative agents has not been determined. With our present knowledge it is impossible to explain either the spontaneous origin of these agents within an individual or to account for their transfer from one individual to another under natural conditions. However, the above work on avian neoplasm together with similar work on the smaller mammals gives us some ground to hope that the cause, methods of prevention and cure for tumor and cancer in animals and man may soon be discovered.

CHAPTER XXI

Poultry Surgery

It is proposed to consider in this chapter only those common pathological conditions of poultry which demand surgical treatment for their cure. At the outstart it should be said that poultry bear and recover from surgical operations very well. The common practice of caponizing, usually done without any aseptic precaution whatever and with small losses from infection, is sufficient evidence of this. Probably no mammal would bear opening the abdominal cavity (which is done in every caponizing operation) with such entire and nearly uniform freedom from ill effects as attends this operation with poultry. The reason why poultry make such excellent surgical subjects lies in their marked resistance to all pyogenic (pus producing) germs.

The Treatment of Cuts, Tears and All Open Wounds

Very severe wounds may be successfully treated by adhering to the following procedure:

1. Thoroughly wash the hands in warm water, using plenty of soap, before handling the wounds at all. After the hands have been well scrubbed, rinse them thoroughly in a pan of 1 to 1000 bichloride of mercury solution (p. 54) and dry with a *clean* towel.

2. Pull out the feathers in the region around the wound, and thoroughly cleanse it, using *first warm water*, and follow this with *warm 1 to 1000 bichloride solution*. A piece of clean

soft cloth may be used for this purpose, or absorbent cotton. Make sure that the wound is *thoroughly* clean. Do not be afraid of hurting the bird. A little pain at the start is preferable to a dead bird later.

3. If necessary sew up the wound, using a good sized sewing needle and silk. Both needle and silk should be soaked in alcohol for 15 minutes before using. Small wounds need not be sewed. Large ones will heal much quicker and more certainly if they are sewed. If the wound involves the muscles as well as the skin sew it up in two layers; one set of stitches including only the muscles, the other set only the skin.

4. Paint the skin in the region about the wound, *but not the wound itself* with dilute tincture of iodine.

5. Powder the wound well with iodoform.

6. Smear a thick layer of the ointment already recommended (p. 55) over all.

7. If the wound is very severe bandage it with a clean cloth.

The above treatment is only necessary in its entirety in very severe cases. Depending upon the gravity of the condition the following items in the treatment may be omitted in the order named:

7 may be omitted except in most serious cases.

7 and 3 may be omitted in less severe cases.

7, 3 and 4 may be omitted in still less severe cases.

7, 3, 4, and 1 may be omitted in still less severe cases.

In case of slight wounds which appear still to demand some treatment 6 and 2 or even 6 alone will suffice.

Abscess

Should an abscess appear lance it with a clean sharp knife, making sure to cut to the bottom. Squeeze out the pus and core if there is one, and then proceed to heal it by following the treatment above outlined for wounds in general.

Bumblefoot

This is an abscess of the foot which may result from a variety of causes, *e.g.*, too high roosts, too narrow roosts, undiscovered wounds caused by stepping on nails, splinters of glass, etc. It is usually not discovered until the bird becomes lame.

The best treatment to follow is first to tie a cord *tightly* about the leg above the foot to control the flow of blood; then with a *clean*, narrow bladed, *sharp* knife open up the abscess *thoroughly*. Go clear to the bottom and dig out the core. Then follow *in detail, omitting nothing except 3*, the treatment given above for wounds. Two days after the first treatment take off the bandages and repeat the treatment, going through in order steps, 1, 2, 4, 5, 6, and 7. In some cases a third treatment after a lapse of 2 or 3 days may be necessary, but usually not if the first treatment is thorough.

Of course the bird under treatment should be isolated and kept in a small pen with soft litter on the floor.

Broken Bones

If a bird is sufficiently valuable to warrant the trouble it is possible to set fractures of the long bones of legs and wings, and get successful union. A splint should be made for the affected part and carefully and thoroughly bound into place. Healing is rapid, and it should be possible to remove the splints in three weeks from the time they are put on if not before. In our experience firm union has occurred in less time than this.

Frozen Combs and Wattles

In northern parts of the country frozen wattles and combs, especially in male birds, are very common occurrences. The trouble is more apt to be with the wattles than the comb, because the former dip into the drinking water and then freeze at times when if dry they would not do so.

The following brief but adequate directions for treating frozen combs and wattles are taken from *Farm Poultry*, Vol. 15, p. 41: "First thaw the wattles or combs out by manipulating with the fingers well smeared with vaseline. Keep the bird in a cool (not cold) place, and anoint the frozen parts with a mixture of vaseline, 5 tablespoonfuls; glycerin, 2 tablespoonfuls; turpentine, one tablespoonful, once or twice a day. If he is not very badly frosted it probably will make no difference with his breeding a few months from now — provided he is not again injured the same way."

Anæsthetizing Poultry

The difficulty which we have found to be inherent in anæsthetizing the domestic fowl may be stated briefly in this way: If any anæsthetic is pushed to the point at which the bird is in satisfactory condition for operative procedure in about 9 cases out of 10 the bird will die on the table from the effects of the anæsthesia before the operation, if extensive, can be completed. If, on the other hand, the anæsthetic is given less freely the bird does not lose its reflex excitability. Every time a cut is made or a nerve is pinched with the forceps the bird will struggle. Our experience in anæsthetizing birds, which has now covered a large number of individuals, leads us to believe that the only middle ground between these two extremes is afforded by those cases (unfortunately too few) in which the individual idiosyncrasy of the bird toward ether makes it take the anæsthetic well.

While we have made no detailed physiologic study as to the fundamental reasons underlying this difficulty respecting anæsthesia which has been described, it seems reasonably apparent what these reasons are. Connected with the respiratory organs proper of a bird are the relatively enormous air sacs. During anæsthesia the ether or chloroform vapor gets into these air sacs either by diffusion or directly as a result of respiratory movements. There is reason to believe that the vapor, once in the air sacs, stays there until it is absorbed by the tissues; in other words, it appears to be the case that the great bulk of an inhaled anæsthetic in the case of birds must be eliminated from the body by way of the urinary organs rather than the respiratory organs. Assuming this to be the case there is no difficulty in seeing why forcing an anæsthetic in a bird leads to disastrous results. The relatively enormous area for absorption afforded by the air sacs insures that a correspondingly large amount of the anæsthetic will be taken up very quickly. This almost immediately affects the vagus center, with the consequent cardiac inhibition, respiratory failure and death.

The exact method of procedure which we now follow in anæsthetizing birds is as follows: Immediately before beginning the administration of the anæsthetic a 1–200 grain atropin sulphate tablet is dissolved in 1 c.c. of warm normal saline solution. The salt solution with the dissolved atropin is then injected subcutaneously in the axilla. Ether is used as the anæsthetic. It is administered from a small improvised mask which admits of the condition of the comb being seen during the operation. Depending on how hard the ether is pushed, the bird is ready for operation in from 15 to 20 minutes after the anæsthesia is begun. The dosage of 1–200 grain atropin to a bird may seem large, but we have never been able to see the slightest bad effect from it, provided the administration of ether was begun immediately after the injection of the atropin.[1]

[1] The foregoing account of anæsthesia is taken from a paper by two authors (Pearl and Surface) published in *Jour. Amer. Med Assoc.*, Vol. 52, pp. 382–383, 1909.

GLOSSARY OF TECHNICAL TERMS

Abdomen. — That portion of the body which contains the internal organs. Belly.

Agglutination. — The massing together of bacteria or the red corpuscles of the blood when subjected to certain substances.

Air sac. — One of the membranous sacs filled with air in different parts of the body, especially in the abdominal region. They often extend into the cavities of the bones and connect with the lungs.

Albumen portion of oviduct. — See p. 248.

Amœbiasis. — A diseased state produced by amœbæ.

Anæmia. — A condition in which the blood is deficient either in quality or quantity. It is marked by paleness and loss of energy.

Anterior. — Situated in front of or in the forward part of.

Anus. — The external opening of the intestine. Vent.

Apathetic. — Lacking in feeling or ambition. Indifferent.

Arachnida. — A class of invertebrate animals including among other groups the spiders, scorpions and mites.

Articular. — Pertaining to the joints.

Astringent. — Causing contraction and arresting discharges.

Atony. — Lack of normal tone or strength.

Atrophy. — A wasting or diminution of the size of a part.

Auditory meatus. — The opening into the ear.

Autopsy. — The post-mortem examination of a body.

Avian. — Pertaining to birds.

Axilla. — The region under the wing where the latter joins the body.

Bile. — The substance secreted by the liver. Gall.

Bronchi. — The tubes which lead from the end of the windpipe (trachea) to the two lungs (cf. Fig. 26).

Carcinoma. — A malignant tumor or cancer.

Catheter. — A tubular surgical instrument for discharging fluids from a cavity of the body or for distending a passage.

Cecum (pl. ceca). — A blind intestinal pouch of which there are two in the fowl.

Cell. — The smallest element of an organized body that manifests independent vital activities. A morphological or structural unit of an organism.

Chronic. — Long continued but not acute.

Cleavage. — The division of the cells of an embryo.

Cloaca. — The enlarged portion of the alimentary canal just before the vent. The intestine, the ureters (tubes from the kidneys) and the oviduct open into the cloaca.

Concrement. — A lump or mass formed by successive additions.

Congestion. — Excessive or abnormal accumulation of blood in a part or organ.

Conjunctiva. — The delicate membrane that lines the eyelids and covers the eyeball in front.

Contagious. — A disease which is communicable by direct contact.

Copulation. — Sexual intercourse. With fowls "treading."

Cornea. — The hard transparent structure forming the anterior part of the eyeball.

Cranium. — The brain case.

Creolin. — A thick black liquid coal tar preparation. It has antiseptic properties.

Cresol. — A coal tar product with antiseptic and germicidal properties.

Cyst. — A sac-like growth which usually contains a liquid or a semisolid.

Demulcent. — A soothing mucilaginous or oily medicine.

Dermoid cyst. — A form of congenital cyst often containing skin-like structures.

Diuretic. — A medicine that increases the activity of the kidneys.

Dorsal. — Pertaining to the back.

Ecchymoses (ek-kim-o-ses). — Discoloration of the skin caused by blood outside of the blood vessels as in a bruise.

Emaciated. — Very lean or wasted condition of the body.

Enema. — A liquid injection in the rectum or cloaca.

Enteritis. — Inflammation of the intestine. In human medicine confined chiefly to the small intestine.

Epidemic. — A disease that is widely prevalent in a community or locality.

Epidermis. — The outer or non-vascular layer of the skin. The cuticle.

Epithelioma. — A cancer or malignant tumor consisting chiefly of cells derived from the skin or mucous membrane.

Epithelium. — The covering or outer layer of the skin and mucous membranes.

Epizoötic. — Occurring as an epidemic disease among animals.

Ergot. — A fungus which affects and finally replaces the seed of a cereal grass. Used chiefly in connection with the ergot of rye, which is poisonous to poultry. Ergot as a drug has the property of causing the mammalian uterus to contract.

Etiology. — The causation of any disease.

Excrement. — Fecal matter; matter cast out as waste from the body.

Exudate. — A substance thrown out of the body or deposited in a tissue by a vital process.

Feces. — The excrement or undigested residue of the food discharged from the intestines. Dung. Droppings.

Flagellate micro-organism. — Any minute microscopic organism which swims through the water by means of the lashing of one or more hair-like structures (flagella).

Follicle. — See p. 246.

Gall bladder. — The reservoir for the bile or gall secreted by the liver. It is readily seen on the upper side of the liver.

Gallus domesticus. — The scientific name for the domestic fowl.

Gangrene. — See p. 271. Gangrene of oviduct.

Gastritis. — Inflammation of the stomach.

Hermaphrodite. — An organism which has both male and female reproductive organs.

Hemorrhage. — Bleeding. A copious escape of blood from the vessels.

Hepatic. — Pertaining to the liver.

Hyperæmia. — Excess of blood in any part of the body.

Hypertrophy. — The morbid enlargement or overgrowth of an organ or part.

Immunity. — Security against any particular disease.

Infection. — The transmission of disease from one animal to another, usually through some intermediate agent.

Impaction. — The condition of being firmly lodged or wedged.

Incinerate. — To burn to ashes. Cremate.

Incision. — A cut.

Infiltration. — The accumulation in a tissue of substances not normally found in it.

Inflammation. — A morbid condition characterized by pain, heat, redness, and swelling, and by hyperæmia and various exudations.

Inoculation. — The insertion of a virus into a wound or abrasion in the skin in order to communicate a disease.

Isthmus. — See p. 248.

Keratitis. — Inflammation of the cornea of the eye.

Larva. — The first stage in development after leaving the egg. Used in connection with insects, worms, etc.

Larynx. — A muscular and cartilaginous structure situated at the base of the tongue and connecting with the windpipe (trachea). It is the organ of voice.

Lesion. — Any hurt, wound or local degeneration.

Leucocytes. — White blood corpuscles.

Lumen. — A transverse section of the clear space within a tube.

Lymphatic. — Pertaining to or containing lymph which is a transparent slightly yellow liquid which fills the lymphatic vessels. It corresponds in some respects to the serum or liquid portion of the blood.

Mammal. — Any vertebrate animal which suckles its young.

Melanosis. — Pertaining to an abnormal deposit of pigment.

Mesentery. — The fold of peritoneum attached to the intestines.

Metamorphosis. — In insects the change from larval to adult form as from caterpillar to butterfly.

Micro-organism. — Any minute (microscopic) animal or plant. Often used in referring to bacteria or germs.

Mite. — A small arthropod somewhat related to spiders (cf. Fig. 40).

Morphological. — Pertaining to the forms and structures of organized beings.

Mucosa. — The mucous membrane.

Mucous membrane. — The lining of the internal cavities of the body.

Mucus. — The viscid secretion of certain (mucous) glands.

Mycelium. — The thread-like portion of a fungus (cf. Fig. 53).

Nacreous. — Resembling mother-of-pearl.

Necrosis. — Death of a tissue.

Necrotic. — Pertaining to dead or decaying tissue.

Nictitating membrane. — The third or lateral eyelid in birds and some related forms, springing from the inner or anterior border of the eye and capable of being drawn across the eyeball.

Nucleus (pl. nuclei). — A spherical body within a cell. The nucleus is essential to the life of the cell.

Œsophagus. — That portion of the alimentary canal between the mouth (pharynx) and the crop.

Oral. — Pertaining to the mouth.

Ovary. — The female sexual organ in which the eggs develop.

Oviduct. — The tube through which the egg passes from the ovary to the cloaca.

Ovum (pl. ova). — The egg, particularly while in the ovary, (cf. Fig. 60).

Panophthalmia. — Inflammation of all the structures or tissue of the eye.

Papilla. — A small nipple shaped elevation.

Paralysis. — A loss of motion or sensation in a living part or member.

Pathology. — That branch of medicine which treats especially of the tissue changes caused by disease.

Pectoral. — Pertaining to the breast or chest.

Pelvis. — The girdle or ring of bone at the posterior extremity of the trunk, supporting the spinal column and resting upon the legs.

Pericardium. — The membranous sac which contains the heart.

Peristalsis. — The worm-like movements of the intestine and oviduct by which the contents of these tubes are propelled.

Peritonitis. — Inflammation of the peritoneum or the membrane lining the abdominal cavity.

Pharynx. — That portion of the alimentary canal between the mouth and the œsophagus. It also communicates with the larynx and nasal passages at its upper end.

Pleural. — Pertaining to the serous membrane which covers the lungs.

Posterior. — Situated behind or towards the rear.

Post-mortem. — Latin for *after death.* See autopsy.

Prognosis. — The prospect as to recovery from a disease or a forecast as to the probable result of an attack of a disease.

Protoplasm. — A viscid granular material which forms the essential constituent of the living cell. Living substance.

Protozoa. — A class of unicellular animal micro-organisms.

Proventriculus. — That portion of a bird's alimentary canal lying between the crop and the gizzard. Often called the stomach.

Punctiform hemorrhages. — Presenting the appearance as if punctured by a large number of fine prickle or needle holes from which the blood oozes.

Purgative. — Causing evacuations of the bowels.

Pyæmia. — Blood poison due to microbic origin.

Sarcoma. — A kind of tumor or cancer not always of a malignant nature.

Scabies. — A contagious skin disease caused by a mite.

Sclerotic. — Pertaining to the hard white fibrous membrane which with cornea forms the outermost coats of the eyeball.

Serum. — The clear liquid which separates from the clot and the corpuscles in the clotting of blood.

Spleen. — An oval shaped organ normally about one-half inch in diameter and of a dark red color. It lies immediately above the liver and between that and the proventriculus.

Spirochete. — A protozoön parasite belonging to the genus *Spirochæta.*

Spore. — The reproductive cell of many protozoa and of many lower plants. It is usually inclosed in tough membranes and is difficult to kill.

Stigma. — See p. 248.

Subcutaneous. — Beneath the skin.

Sub-mucosa. — The layer of tissue situated beneath the mucous membrane.

Syncope (sin-ko-pe). — Fainting. Failure of the heart's action.

Trachea. — The wind-pipe.

Traumatic. — Caused by an injury.

Therapeutic. — Pertaining to the art and science of healing.

Thoracic. — Pertaining to the chest.

Urate. — A salt of uric acid. A product of the secretion of the kidneys. The white part of a fowl's droppings.

Ureters. — The tubes leading from the kidneys to the cloaca.

Uterus. — See p. 248.

Vagina. — That portion of the oviduct between the shell gland and the cloaca.

Ventral. — Pertaining to, or situated toward, the belly.

Virulent. — Extremely poisonous or dangerous.

Virus. — Any animal poison, especially one produced by and capable of transmitting a disease.

Viscera. — The internal organs of the body.

INDEX

Abdomen, baggy, 80; liquid in, 80; swollen, 79.
Abdominal dropsy, 80.
Abnormal eggs, 272.
Abortion of eggs, 256.
Abscess, 325.
Acarina, 213.
Achorion schonleinii, 234.
Aconite root, 55.
Adenoma, 320, 322.
Agglutination test for tuberculosis, 127; for white diarrhea, 297.
Air, 19.
Air-sac mite, 180, 227.
Air sacs, 147.
Ajowan oil, 169.
Alimentary tract, diseases of, 57.
Amœba meleagridis, 95.
Anæmia, 120.
Anæsthetizing poultry, 327.
Anatomy, of the fowl, 41; of reproductive organs, 245; of respiratory organs, 147.
Apoplectiform septicæmia, 189.
Apoplexy, 194.
Apothecaries' weights and measures, 56.
Argas persicus, 192, 228.
Arsenic as poison, 82.
Articular gout, 200.
Ascaris inflexa, 144.
Ascites, 80.
Aspergillosis, 101, 173; in chicks, 302.
Aspergillus species, 175; *fumigatus*, 176, 285, 303.
Atoxyl, 193.
Atrophy of liver, 93; ovary, 251.
Autopsy, 46.

Bacillary white diarrhea, 287.
Bacillus avisepticus, 103; *bipolaris septicus*, 103; *fluorescens liquefaciens*, 321; *pullorum*, 112, 285, 287, 291, 295; *suisepticus*, 104; *tuberculosis*, 116; *typhi gallinarum alcalifaciens*, 112; *typhosis*, 111.
Bacterium sanguinarium, 11.
Baldness, 233.
Balfour, 191.
Bang, B., 131.
Bang, Oluf, 118.
Banks, 206, 207.
"Bed-bug" of poultry, 230.
Beeck, 82.
Beri-beri, 197.
Berke, 146.
von Betegh, 156, 160.
Bichloride of mercury, 54, 324.
Blackhead, 94.
Blood, diseases of, 185.
Blood vessels, rupture of, 184.
Bloody diarrhea, 68, 83, 91, 138.
Bloody spots in eggs, 278.
Body mange, 226.
Bollinger, 183.
Bones, broken, 326.
Boracic acid, 152.
Bordet et Fally, 160.
Borrel, 160.
Bradshaw, 52, 142.
Brain, congestion of, 195; hemorrhage of, 194; post-mortem appearance of, 194.
Breaking of egg in oviduct, 271.
Breast bone, crooked, 310.
Breeding for health and vigor, 3.
Broken bones, 326.
Bronchi, 147.
Bronchitis, 153.
Brown, 27.
Bruet, 240.
Bumblefoot, 326.
Burckhardt, 186.

Calomel, 52.
Cancers, 101.
Canker, 164, 166, 172.
Carbolic acid for cholera, 109; for catarrh, 152.
Carcinoma, 320.
Carcinomatosis, 101.
Carnwath, 156.
Cary, 155, 240, 243.
Castor oil, 53.
Catarrh, simple, 151; severe, 153; contagious, 156.
Catechu, 52.
Cayenne, 52.
Ceca, post-mortem appearance of, 97, 98, 294.
Cephalogonimus pellucidus, 145.
Cercomoniasis, 100.
Cerebral hyperæmia, 195.
Chelosperura hamulosa, 143.
Chicken pox, 155, 237.
Chickens, diseases of, 301.
Chilomastix gallinarum, 146.
Cholera, 103, 188.
Circulatory system, diseases of, 182.
Cleanliness, 14.
Cloacitis, 280.
Coccidia, 96.
Coccidiosis, 71, 285.
Coccidium tenellum, 285; cuniculi, 285; life history of, 73.
Cock eggs, 273.
Cold, 151; treatment for, 152.
Cole and Hadley, 75, 95, 96, 285.
Comb, frozen, 326; white, 233, 236.
Common measure, equivalents of, 56.
Complement fixation, 127.
Congestion of lungs, 177; brain, 195.
Constipation, 69, 70.
Constitutional vigor, breeding for, 3; definition, 6.
Convulsions, 83, 196.
Copper poisoning, 83.
Corrosive sublimate, 54.
Cotton seed oil, 54.
Cremation, 31.
Crematory, 31, 33.
Creolin, 152.
Cresol soap, 17.
Cresol solution, 15; disinfectant, 17.

Crooked breast bone, 310.
Crop, impacted (crop bound), 58; inflammation of, 61; catarrh of, 61; enlarged, 62; inflated, 63; paralyzed, 105.
Crop, post-mortem appearance of, 48, 294.
Cropping poultry ranges, 27.
Croup, 153.
Crurea, 143.
Curtain-front house, 11; interior, 13.
Curtice, 102.
Curtis, 78, 258, 274.
Cuts, treatment of, 324.
Cysticercoid, 136.
Cytodites nudus, 180, 181.

Dammann and Manegold, 190.
Dampness, 10, 21.
Dandelion, for liver trouble, 92.
Davainea proglottina, 142.
Davainea tetragona, 137.
Dawson, 188.
Dead birds, disposal of, 30.
Decayed food, 85.
Denny, 207.
Depluming scabies, 225.
Depperich, 113.
Dermanyssus gallinæ, 214.
Diagnosis of disease, 36.
Diagnosis, table for differential, 37, 50.
Diarrhea, 64, 67, 70, 85, 98, 105, 120, 154, 167, 174; bacillary, 287; bloody, 85, 91, 167; diagnosis of, 292; watery, 89; white, 95, 112, 283.
Digestive organs, inflammation of, 106.
Diphtheria, avian, 155, 164; vaccine for, 169; serum for, 169.
Diphtheritic roup, 164, 166.
Disease, prevention of, 3.
Disinfection, 15; formalin, 16; formaldehyde gas, 16; cresol, 17.
Dispharagus spiralis, 143; *nasutus*, 143.
Dissection of bird, 46.
Distemper, 154.
Distoma ovatum, 145; species, 145.

Double-yolked eggs, 274.
Dove cot bug, 230.
Drepanidotænia infundibuliformis, 135.
Drinking water, 24; antiseptic for, 25.
Droppings, green, 105; normal, 65.
Dropsy, 80.
von Durski, 257, 279.
Dysentery, 67.
Dyspepsia, 70.

Edema of wattles, 244.
Egg, bound, 266; breaking of, in oviduct, 271; laying, physiology of, 249.
Eggs, abnormal, 272; abortion of, 256; soft-shelled, 273; yolkless, 273; "cock," 273; "witch," 273; double-yolked, 274; triple-yolked, 274; inclusions in, 278; spots in 278; small, 273.
Ehrenreich, 313, 320.
Ehrenreich and Michaelis, 313.
Ellerman and Bang, 186.
Emaciation, 120.
Emphysema, 304.
Endocarditis, 183.
Enlargement of heart, 184; liver, 90.
Enteritis, 67.
Enterohepatitis, infectious, 94.
Epidermoptes bilobatus, 226; *bifurcatus*, 226.
Epilepsy, 196.
Epithelioma contagiosum, 155, 237.
Epizoötic, 154.
Epsom salts, 53.
Ergot of rye, 84.
Eversion of oviduct, 263.
Exercise, 29.
External parasites, 30, 203; keeping poultry free, 203.
Extractor, gape worm, 310.
Eye worm, 232.
Eyes, roup of, 157, 162.

Fally, 156.
Fantham, 75.
Fatty degeneration of liver, 92.
Favus, 226, 233.

Feeding, hygienic, 21.
Fleas, 230, 231.
Flukes, 145.
Follicle, 246; failure to rupture, 257.
Formaldehyde gas, 16.
Formalin, 16.
Fowl cholera, 102, 188.
Fowl plague, 112.
Fowl typhoid, 102, 111, 186.
Freese, 113.
Freidberger and Frohner, 237, 241.
Fresh air, 10, 19.
Frozen comb and wattles, 326.
Funk, 197.

Gadow, 149.
Gage, 288, 297.
Gage and Opperman, 53, 139, 140, 141.
Galli-Valerio, 160.
Game, Cornish Indian, 251.
Gangrene of oviduct, 271.
Gapes, 304.
Gastritis, 63.
Geese, 113.
Gerhartz, 249.
Gingylonema ingluvicola, 143.
Gizzard, post-mortem appearance of, 48.
Gleet, vent, 280.
Glossary, 329.
Going light, 173.
Goniodes dissimilis, 207.
Gonococcus, 281.
Gout, 101, 199; visceral, 200; articular, 200.
Greene, 85, 91, 269.
Green droppings, 105.
Green food, 24.
Grippe, 154.
di Gristiana, 201.
Guerin, 160.
Guinea-fowl, 113.
Gurlt, 257.

Hadley, 105, 109, 110.
Hadley and Amison, 110.
Hadley and Beach, 155, 239, 243.
Hadley and Kirkpatrick, 286.
Haiduk, 218, 219, 220, 221, 223.
Hamilton, 253.

Haring and Kofoid, 156.
Harrison, 240.
Harrison and Streit, 155, 158, 160, 162.
Harvest-bug, 227.
Hauer, 193.
Health, breeding for, 3.
Health type, 4.
Heart, diseases of, 182; enlargement of, 184; hypertrophy of, 184; rupture of, 184.
Heart, post-mortem appearance of, 47, 106, 113, 182, 183, 188, 200.
Heart sac, dropsy of, 182.
Heat prostrations, 195.
Hebrant and Antoine, 200.
Helodrilas parvus, 143.
Hermaphroditism, 252.
Hemorrhage, of brain, 194.
Heterakis perspicillum, 143, 144.
Higgins, 115.
Hill, 182, 222.
Himmelberger, 119, 127.
Hirschfeld and Jacoby, 185.
Horton, 300.
Housing, poultry, 9.
Hydrogen peroxide, 152.
Hygiene, 8; essentials of, 35.
Hyperæmia, cerebral, 195.
Hypertrophy of heart, 184; of liver, 90; of yolk, 257.

Illness, symptoms of, 37.
Immunity against cholera, 109.
Inclusions in eggs, 278.
Indigestion, 70.
Infectious enterohepatitis, 94.
Infectious leukæmia, 111, 112, 185.
Inflammation of mouth, 171; of oviduct, 262.
Influenza, 154.
Internal parasites, 133.
Inoculation for cholera, 109; for roup, 169.
Intestinal cocciciosis, 285.
Intestinal worms, 133, 196.
Intestines, congestion of, 86.
Intestines, post-mortem appearance of, 48, 68, 86, 87, 106, 123, 124, 294, 303.
Isolation of sick birds, 34.

Jaundice, 94.
Jewett, 75.
Johne's disease, 119.
Jones, 288, 297.

von Katz, 156, 160.
Kaupp, 231, 300.
Kidneys, congestion of, 86; diseases of, 199; enlarged, 199; inflammation of, 106; post-mortem appearance of, 49, 86, 122, 199, 294.
King and Hoffman, 113.
Kingsley, 155.
Kitt, 110.
Knemidocoptes (*Dermatoryctes*) (*Sarcoptes*) *mutans*, 216, 218, 220, 221.
Koch, 320.
Koch and Rabinowitsch, 119, 122, 128.
Kolle and Hetsch, 191, 192.

Lameness, 120, 121.
Land, 26.
Landois, 279.
Landsterner, 113.
Larynx, 147.
Laurie, 228, 229.
Lawry, 211.
Lead as poison, 83.
Leg weakness, 301.
Leukæmia, 185; infectious, 111, 112, 186.
Levaditi and Manouclian, 191.
Levaditi and McIntosh, 193.
Lewis, 5.
Lewis and Clark, 281.
Lice, 206; life history of, 207; mercurial ointment for, 205; method of infestation, 208.
Lice powder, how to make, 211.
Life history of coccidium, 73.
Light, 19.
Limberneck, 86, 199, 202.
Lipeurus heterographus, 207; *variabilis*, 207.
Lisoff, 110.
Litter, 20.
Liver, congestion of, 86, 88, 93; diseases of, 87; nodules on, 87; spotted 87, 92, 95; hypertrophy,

90; enlarged, 90; fatty degeneration of, 92; atrophy of, 93; post-mortem appearance of, 47, 68, 86, 87, 92, 93, 94, 95, 98, 106, 122, 125, 187, 294, 303.
Liver disease, 87; cause of, 88; treatment of, 89.
Loùnoy and Bruhl, 191.
Lowenstein, 129.
Lungs, congestion of, 177; post-mortem appearance of, 48, 122, 125, 147, 294.
Lye as poison, 81.

Maine Experiment Station, 11, 17, 24, 25, 27, 31, 164, 204, 211.
Male reproductive organs, diseases of, 282.
Mallophaga, 206.
Mange, 226.
Manson's eye worm, 232.
Manteufel, 243.
Marchoux, 114.
Martin and Robertson, 146.
Materia medica, 52.
Measures, 56.
Medical Record, 129.
Medicines, 52.
Mégnin, 94, 217, 235, 305, 307.
Menopon biseriatum, 206; *pallidum*, 206, 208.
Mercurial ointment, 205.
Mercury, bichloride of, 54, 324.
Mercury poisoning, 83, 205.
Mesogonimus commutataris, 145.
Metchnikoff, 300.
Metric equivalents, 56.
Meyer and Crocker, 76.
Mitchell and Bloomer, 111.
Mites, 114, 213; air-sac, 180, 227; connective tissue, 227; deplumimg, 224; harvest-bug, 227; red, 214.
Mohler and Buckley, 176.
Monocercomonas gallinarum, 100.
Moore, 102, 111, 137, 186, 188, 189.
Moore and Ward, 115.
Morse, 25, 131, 285, 286.
Mouth, inflammation of, 171.
Muller, R., 160.
Müller, J., 107.

Murray, 227.
Mycosis, 173.
Myocarditis diphtheritica, 183.

Nematode worms, 142.
Nervous system, diseases of, 194.
New flock building, 131.
New Jersey Experiment Station, 5.
Neumann, 145.
Nitrate of soda as poison, 81.
Nits, 208.
Nodular tæniasis, 137, 138.
Notocotyle triserialis, 145.

Obstruction of oviduct, 266; of vent, 69.
Ointment, for wounds, 55; mercurial, 205.
Operation for egg bound, 269; impacted crop, 60.
Oppel, 150.
Oregon Experiment Station, 300.
Osborn, H., 214.
Ostertag and Ackermann, 106.
Ovary, 246; atrophy of, 251; diseases of, 251; gangrene of, 255; tumors on, 256.
Ovary, post-mortem appearance of, 48, 122.
Overfeeding, 24.
Oviduct, 248, 260; anatomy of, 258; diseases of, 258; inflammation of, 262; prolapse of, 263; obstruction of, 266; rupture of, 270; gangrene of, 271; broken egg in, 271.
Ovule, 246.
Oxysperura mansoni, 232.

Paralysis, partial, 85.
Parasites, external, 203; internal, 133.
Parrots, 119, 121.
Pasteur, 104.
Pearl and Curtis, 78, 250.
Pearl and Surface, 328.
Pearson, 196, 233.
Pearson and Warren, 138, 195.
Pericarditis, 182.
Pericardium, inflammation of, 182.

Peritoneum, post-mortem appearance of, 49, 79.
Peritonitis, 77.
Permanganate, potassium, 25, 152; for roup, 163.
Pernot, 115, 120.
Pfeiler and Rehse, 112.
Pharynx, 147.
Pheasants, 113, 253, 254.
Philips, 170.
Phosphorus poisoning, 84, 113.
Physiology, of reproductive organs, 245; respiratory organs, 147.
Pick, 320.
Pickens, 320, 322.
Pierce, 216.
Pip, 171.
Plague, fowl, 102, 112.
Plymouth Rock, 251.
Pneumomycosis, 302.
Pneumonia, 178.
Poisons, 81; ptomaine, 85; treatment for, 86.
Poisonous plants, 85.
Polish, White Crested Black, 251.
Polyneuritis, 197.
Post-mortem appearance of brain, 194; ceca, 97, 98, 294; heart, 47, 106, 113, 182, 183, 188, 200; intestines, 48, 68, 86, 87, 106, 123, 124, 294, 303; kidneys, 49, 86, 122, 199, 294; liver, 47, 68, 86, 87, 92, 93, 94, 95, 98, 106, 122, 125, 187, 294, 303; lungs, 48, 122, 125, 147, 174, 177, 178, 183, 188, 294, 303; ovary, 48, 122; spleen, 48, 68, 87, 122, 125, 294.
Post-mortem appearances, 47, 48; diagnostic value of, 49; table of, 50.
Post-mortem examination, 40; directions for making, 46.
Potassium permanganate, 25, 152; for roup, 163.
Poultry surgery, 324.
Powder, lice, 211.
Prevention of cholera, 107; of disease, 3.
Prolapse of oviduct, 263.
Prowazek, 192.

Ptomaine poisoning, 85.
Ptycholes coptica, 169.
Pulex gallinæ, 230, 231.

Rabies, 114.
Range sanitation, 26.
Ransom, 136, 143, 232.
Rats, 156.
Red mite, 214.
Reidenbach, 169.
Reproductive organs, 247; anatomy and physiology, 245; diseases of, 245; diseases of male, 282.
Respiratory system, anatomy and physiology, 147; diseases of, 147.
Rettger and Harvey, 287.
Rettger and Kirkpatrick, 288.
Rettger and Stoneburn, 291, 292, 293, 294.
Rettger, Kirkpatrick and Jones, 289, 290, 291, 295, 297, 299.
Rheumatism, 199, 201; in tuberculosis, 121.
Rhode Island Agricultural Experiment Station, 109, 110.
Robinson, 90, 224, 236, 265, 266, 301.
Roebuck, 157, 166.
Rosenthal, 114.
Rotation, crops and chickens, 27.
Rottiger, 311.
Round worms, 142.
Roup, 155; nasal, 156; diphtheritic, 164.
Rous, 314, 316, 321.
Rous and Lange, 314, 317.
Rous and Murphy, 314, 318, 319.
Rous, Murphy and Tytler, 313, 314, 318, 321.
Rupture, of blood vessels, 184; heart, 184; oviduct, 270.
Russ, 112.

Saccharomyces albicans, 173.
Salmon, 4, 77, 96, 102, 115, 129, 139, 144, 154, 156, 182, 183, 188, 196, 209, 226, 255, 265, 266, 280, 301.
Salt, as poison, 81.
Salts, Epsom, 53.
Salvarsan (606), 193.
Sanitation, 9.

Sarcomatosis, 101.
Sarcoptes mutans, 240.
Sarcoptes lævis var. *gallinæ*, 224, 225.
Scabies, depluming, 224.
Scaly leg, 216.
Schiffmann, 114.
Schmid, 156.
Schalze, 150.
Scott, 143.
Seddon, 244.
Serum for diphtheria, 169.
Sickness, isolation, 34.
Sigwart, 156, 169.
Skeleton, 41.
Skin, diseases of, 233.
Sleepy disease, 189.
Small eggs, 273.
Smith, 95, 96, 286.
Smith and Ten Broeck, 111, 112, 291.
Soft-shelled eggs, 273.
Sore-head, 237, 239.
Sour milk for white diarrhea, 299.
Sparrows, 119.
Spirochæta gallinarum, 191.
Spirochæta marchouxi, 230.
Spirochætosis, 190.
Spleen, post-mortem appearance of, 48, 68, 87, 122, 125, 294.
"Spotted liver," 100.
Spots in eggs, 278.
Staggering, 195.
Stieda, 149.
Stiles, 135, 136, 138, 142.
Stock tonic, 71.
Stomach, inflammation of, 63.
Storrs Experiment Station, 288.
Streit, 160.
Streptococcus capsulatus gallinarum, 190.
Strychnine, 84.
Suffram, 81.
Sunlight, 10, 19.
Surface, 259.
Surgery, poultry, 324.
Sweet, 156, 240.
Symplectoptes cysticola, 227.
Symptoms, table of external, 37.
Syngamus trachealis, 305, 306, 307.

Table of post-mortem appearances, 50.
Tables of symptoms, 37.
Tablets, 54.
Tæniasis, nodular, 137.
Tainted ground, 27.
Tape worms, 134; treatment, 139; prevention, 141.
Tears, treatment of, 324.
Tetranychus (Leptus) autumnalis, 227.
Theobald, 145, 181, 225, 227, 228, 232.
Thompson, D. S., 7.
Thompson, J. A., 145.
Thrombosis, 185.
Thrush, 173.
Ticks, 114, 192, 228.
Tonic, 71.
Trachea, 147.
Trematodes, 145.
Trichomastix gallinarum, 146.
Trichomonas gallinarum, 146.
Trichosoma strumosum, 143.
Triple-yolked eggs, 274.
Trypanosoma eberthi, 146.
Tubercle, 122.
Tuberculin, 126; reaction to, 127.
Tuberculosis, 115, 137, 180; diagnosis, 125; method of contagion, 128; treatment for, 129.
Tumors, 101, 121, 312; ovarian, 256; kinds of, 315.
Turkeys, blackhead in, 94, 113.
Typhoid, fowl, 102, 111, 187.
Tyzzer and Ordway, 315, 318, 321.

Uhlenhuth and Gross, 193.
Uhlenhuth and Manteufel, 156, 160.
Ulcers, 121.
United States Department of Agriculture, 216.
Urates, green, 105.
Urates, yellow, 104.

Vaccine for diphtheria, 169.
Vale, 144, 236, 304.
Van Es, 125.
Van Es and Schalk, 119, 123, 125, 126, 127, 130.
Vedder and Williams, 197.

Vent gleet, 280.
Vertigo, 195.
Vigor, breeding for, 3.
Viscera, normal, 43.
Visceral gout, 200.
Vomiting, 61.
Von Linstor and Railliet, 145.

Ward, 106, 108, 122, 124, 128, 240.
Warthin, 185.
Wasting of liver, 93.
Water, drinking, 24.
Wattles, edema of, 244; frozen, 326.
Weights, 56.
White comb, 233, 236.
White diarrhea, 95, 112, 283; bacillary, 287; diagnosis of, 292.
Wilcox and McClelland, 232.

Windpipe, 147.
Witch eggs, 273.
Woods, 293.
Worms, 133; eye, 232; flukes, 145; gape, 305; round, 142; tape, 134; trematode, 145.
Wounds, treatment of, 324.
Wright, 1, 236, 256, 281, 309.
Wry-neck, 202.
Wyandottes, Silver-laced, 7.

Yarrell, 253, 254.
Yolk hypertrophy, 257.
Yolkless eggs, 273.

Zinc as poison, 83.
Zürn, 79, 82, 115, 137, 139, 188, 235, 264, 266.

THE following pages contain advertisements of books by the same author or on kindred subjects

Modes of Research in Genetics

By RAYMOND PEARL, Ph.D.

Biologist of the Maine Agricultural Experiment Station

The field of biological research in which there is to-day the greatest activity is unquestionably genetics. In any new branch of science, little attention is given in the first flush of investigation to the logical concepts and philosophical principles which underlie it. This lack of philosophical poise is now becoming rather generally apparent in genetic research. The present book is a contribution to the methodology of genetics in a philosophical sense. It attempts first to examine carefully and then to appraise the value of the more important current methods of attacking the problems of heredity and breeding, including the statistical or biometrical method, Mendelism, etc. The book should, on the one hand, interest every professional student of biology, in any of its branches, who is at all concerned with the question of the philosophical foundation of his science. On the other hand, the publicist and man of affairs who is concerned to know what significance is to be attached to the eugenics movement should find in this book some aid in orienting himself.

THE MACMILLAN COMPANY

Publishers   64-66 Fifth Avenue   New York

Farm Poultry

By GEORGE C. WATSON, M.S.
Professor of Agriculture in the Pennsylvania State College

New edition revised and rewritten. Illustrated. Cloth, 12mo, $1.50

This is one of the few books designed especially to help the practical farmer in the keeping of poultry. Published originally ten years ago, it contained the gist of the best accepted advice of the day, presenting only those facts that had been proved by experience and which were most capable of application on the farm. The volume has now undergone thorough revision, new ideas and teachings, so far as they safely apply to farm conditions, being incorporated. It is not a fancier's work. The plan of the original has been kept, but the new material in text and pictures is considerable.

Poultry Laboratory Guide

By HARRY R. LEWIS

Cloth, illustrated, 12mo, $0.65

A book which, while primarily intended for use in schools and colleges, presents just the knowledge which will give the amateur hen-keeper a great deal of information without much expenditure of time. The book is illustrated with lucid diagrams which contribute toward the practicability of this work on an interesting and profitable pursuit.

THE MACMILLAN COMPANY
Publishers   **64-66 Fifth Avenue**   **New York**

How to Keep Hens for Profit

By C. S. VALENTINE

Cloth, illustrated, 12mo, $1.50

"The Plymouth Rock, Java, Dominique, Wyandotte, Rhode Island Red, and Buckeye breeds are discussed in the first few chapters. Considerable attention is given to other breeds later on. Eighteen beautiful half-tone engravings adorn the book. From the standpoint of the practical farmer and poultry-grower, we consider this book as one of the very best of its kind. The author is evidently an experienced poultryman. It is a book that should be of special help to beginners in poultry, while at the same time it contains much information for the expert."
—*Farmers' Tribune.*

The Beginner in Poultry

By C. S. VALENTINE

Decorated Cloth, profusely illustrated, 12mo, $1.50

It has been estimated that of the five million people who are raising poultry in this country today half have gone at it blindly. And it is just as impossible to make a success of the poultry business without preparation as it is impossible to succeed in any other business without an acquaintance with the fundamentals. The difficulty which the novice has experienced in going at the raising of chickens systematically in the past has been that he could find no book in which the essentials—only the essentials and all of them—of poultry-raising are given. To write such a book has been Mr. Valentine's purpose. In "The Beginner in Poultry" he discusses the different breeds of fowls, the types of houses, feeding and the kinds of food, raising chickens for the market and for their eggs, diseases and their cures and everything else which will be of value for the one who is starting out—and much for the seasoned poultry-raiser as well.

THE MACMILLAN COMPANY

PUBLISHERS   64-66 Fifth Avenue   NEW YORK

THE RURAL MANUALS   EDITED BY L. H. BAILEY

Manual of Farm Animals
A Practical Guide to the Choosing, Breeding and Keep of Horses, Cattle, Sheep and Swine

By MERRITT W. HARPER
Assistant Professor of Animal Husbandry in the New York State College of Agriculture at Cornell University

Illustrated, decorated cloth, 12mo, 545 pages, index, $2.00

"The work is invaluable as a practical guide in raising farm animals." — *Morning Telegram*.
"A book deserving of close study as well as being handy for reference, and should be in the possession of every farmer interested in stock." — *Rural World*.

Manual of Gardening
A Practical Guide to the Making of Home Grounds and the Growing of Flowers, Fruits, and Vegetables for Home Use

By L. H. BAILEY

Illustrated, cloth, 12mo, 544 pages, $2.00

This new work is a combination and revision of the main parts of two other books by the same author, "Garden Making" and "Practical Garden Book," together with much new material and the result of the experience of ten added years. Among the persons who collaborated in the preparation of the other two books, and whose contributions have been freely used in this one, are C. E. Hunn, a gardener of long experience; Professor Ernest Walker, reared as a commercial florist; Professor L. R. Taft, and Professor F. A. Waugh, well known for their studies and writings on horticultural subjects.

A STANDARD WORK REVISED AND ENLARGED

The Farm and Garden Rule Book

By LIBERTY H. BAILEY

Illustrated, cloth, 12mo, $2.00

When Professor Bailey's "Horticulturist's Rule Book" was published nearly twenty-five years ago, the volume became a standard agricultural work, running through sixteen editions. Taking this book as a basis the author has now made a wholly new book, extending it to cover the field of general farming, stock-raising, dairying, poultry-rearing, horticulture, gardening, forestry, and the like. It is essentially a small cyclopedia of ready rules and references packed full from cover to cover of condensed, meaty information and precepts on almost every leading subject connected with country life.

THE MACMILLAN COMPANY
Publishers   64-66 Fifth Avenue   New York